JOURNAL OF ICT STANDARDIZATION

Volume 2, No. 1 (July 2014)

Special Issue on
Cloud Security and Standardization

Guest Editor:
Monique J. Morrow

JOURNAL OF ICT STANDARDIZATION

Chairperson: Ramjee Prasad, CTIF, Aalborg University, Denmark
Editor-in-Chief: Anand R. Prasad, NEC, Japan
Advisors: Bilel Jamoussi, ITU, Switzerland
Jesper Jerlang, Dansk Standard, Denmark

Objectives

- Bring papers on new developments, innovations and standards to the readers
- Cover pre-development, including technologies with potential of becoming a standard, as well as developed / deployed standards
- Publish on-going work including work with potential of becoming a standard technology
- Publish papers giving explanation of standardization and innovation process and the link between standardization and innovation
- Publish tutorial type papers giving new comers a understanding of standardization and innovation

Aims & Scope

- Aim
  - The aim of this journal is to publish standardized as well as related work making "standards" accessible to a wide public – from practitioners to new comers.
  - The journal aims at publishing in-depth as well as overview work including papers discussing standardization process and those helping new comers to understand how standards work.

- Scope
  - Bring up-to-date information regarding standardization in the field of Information and Communication Technology (ICT) covering all protocol layers and technologies in the field.

JOURNAL OF ICT STANDARDIZATION

Volume 2, No. 1 (July 2014)

Published, sold and distributed by:
River Publishers
Niels Jernes Vej 10
9220 Aalborg Ø
Denmark

www.riverpublishers.com

Journal of ICT Standardization is published three times a year. Publication programme, 2013–2014: Volume 2 (3 issues)

ISSN: 2245-800X (Print Version)
ISSN: 2246-0853 (Online Version)
ISBN: 978-87-93237-12-4

Editorial: Special Issue on Cloud Security and Standardization

Monique J. Morrow

Cisco Services

The industry has been experiencing the deployment of cloud computing models varying from Private Cloud, Hybrid Cloud to an evolution of InterCloud. There are cloud security implications for each model; and within InterCloud a notion of federation may be implied. How do we define cloud bursting and cloud brokering functions and the security requirements therein? What are the use cases? What is the impact of cloud and virtualization on mobile networks and its services? What is the status of standards activities related to cloud security and these scenarios? Authors were invited to submit papers, on these topics for the special issue.

Today's Cloud services are deployed by Cloud Service Providers (CSP) via technologies and business architectures at various places in the network and consumed by customers ranging from large enterprises to consumers. Implicit is the notion of federation amongst CSPs to solve use cases such as disaster recovery, burst demands and geographical coverage, as well as federation between provider and customer, also known as hybrid cloud. Current hybrid-cloud offerings are based on static configurations and static business relationships between a network and a provider or a series of providers. Intercloud will add flexibility to these relationships so that cloud providers could discover services across multiple providers, agree on a common Service Level agreement or create an auction between CSPs to get the lowest price.

Intercloud defines the inter-provider interface (or NNI) between CSPs. In this sense, it is complementary to OpenStack, which focuses on the user-provider interface (or UNI). In order to enable cloud interconnection, a complete architecture is under study, which includes a set of available technologies that will be re-used and some new technologies that are under development. The main enablers for a viable inter-cloud architecture are: the development of a service discovery mechanism amongst multiple providers and customers, enabling multiple cloud provider metadata instantiation, an identity management mechanism between CSPs and real time billing exchange mechanisms. For the industry, as the constructs evolve to hybrid and intercloud implementattions, security AND privacy capabilities will contine to be strategic. Concerns articllated in the industry include loss of control and visibility; service disruption; data security; enterprise isolation and compliance. Cryptography is a fundamental underpinning of nearly all cloud security implementations e.g. emerging homomorphic encryption. Certainly identity management across multiple clouds will also be piovotal.

Various standards bodies have been focusing efforts in the development of cloud security architecture e.g

- **NIST**, http://csrc.nist.gov/publications/nistpubs/800-144/SP800-144.pdf;
- **ITU-T**, http://www.itu.int/ITU-T/newslog/New+ITU+Standards+On+Cloud+Computing+Security+And+Digital+Object+Architecture.aspx;
- **IEEE**, http://cloudcomputing.ieee.org; and so on.

It is with great pleasure that we feature two articles in this special edition:

1. Proposed Identity and Access Management in Future Internet (IAMFI): A Behavioral Modeling Approach by Nancy Ambritta P., Poonam Railkar, Parikshit N. Mahalle, Department of Computer Engineering, Smt. Kashibai Navale College of Engineering, Pune, India
2. Traffic Offload Guideline and Required Year of the 50% Traffic Offloading, by Shozo Komaki, Naoki Ohshima and Hassan Keshavartz. Malaysia-Japan International Institute of Technology, Universiti Teknologi Malaysia, that addresses wireless cloud and implicit security implications.

One can conclude that there is quite a bit of work to do in the cloud security space We will certainly re-visit this topic with best practice guidelines.

With Warm Regards,

Monique J. Morrow
CTO Cisco Services

Proposed Identity and Access Management in Future Internet (IAMFI): A Behavioral Modeling Approach

Nancy Ambritta P., Poonam N. Railkar and Parikshit N. Mahalle

*Department of Computer Engineering, Smt. Kashibai Navale College
of Engineering, University of Pune, Pune, India-411041
Email id: nancy.ambritta@yahoo.com, poonamrailkar@gmail.com,
aalborg.pnm@gmail.com*

Received: April 11, 2014; Accepted: June 19, 2014
Publication: July, 2014

Abstract

The Future Internet (FI) sees the world of objects completely connected over the Internet all the time. It is like opening one's network doors of say home, companies and organizations to the world where it increases efficiency but at any case should not compromise security by exposing sensitive information, presenting tremendous challenge towards access control and identity management in FI. A well-managed identity management system should provide necessary tools for controlling user access and access to critical information. A fitting example will be the IoT (Internet of Things) where every object will be smart and will take advantage of cloud for storage and processing power.

In this paper we provide an introduction to Identity and access management in FI followed by a simplified architecture of the FI and its components. We then proceed by providing a short description about the frequent threats to data stored on cloud along with possible mitigation techniques to the threats. We also provide a comparative study of existing work on access control and propose a method to overcome the limitation of the existing techniques where sensitive organizational information (access policy) is exposed to the cloud. We address this issue in IAMFI by extending the Attribute based encryption

Journal of ICT, Vol. 2_1, 1–36.
doi: 10.13052/jicts2245-800X.211

technique and allowing users to have control over their attribute exposure at the time of requesting access. We also provide a mechanism in IAMFI for distributed attribute and key management for various users thereby reducing the overhead at a single site.

Keywords: IAMFI, Future Internet, Access Control, Identity, Cloud, Attribute-based encryption, Attack Model.

1 Introduction

The Invention of internet is one of the most important discoveries of mankind. It is a network of networks that connects millions of networks through electronic, wireless and optical networking technologies [1]. Billions of people use Internet and it impacts them in almost every aspect of their lives. We cannot imagine a life without Internet today. The original internet architecture was not designed to meet the current demands and the continuous strain in terms of sophisticated threats, performance, reliability and scalability. This has made the research and design of future internet architecture more critical and more relevant than ever before.

Future Internet has never been a single monolithic or single improvement area over the current internet architecture, whereas it is a generic terminology used for the various research activities on internet architecture to address the issues mentioned above and also take into account of bigger picture of the issues [social, ethical, economical] when society is interacting with internet.[2]. Going by the current trends, one example would be moving towards a phase where everything we interact will be connected to the internet. A Fitting example would be Internet of Things where every object will be smart even though it doesn't need to have the necessary processing power as it will take advantage of cloud computing for processing data. In future, not only the phones and computers will be connected whereas every object we use will be connected.

In general, it's like opening one's network doors of say home, companies, and organizations to the world where it increases efficiency but at any case should not compromise security by exposing sensitive information. It presents tremendous challenges towards the access control and identity management in future internet. Hence Identity and access management in future internet needs to be more resilient to attacks, scalable, should maintain privacy and integrity but also on the same hand should have traceability. A well-managed

Identity management system should provide necessary tools and technology for controlling user access and access to critical information.

Thus, we can define Future Internet as a pervasive network of devices connected over the internet that are uniquely identified by their ID's , and interacting with the environment via associated sensors, communicates with the cloud and customizes the environment according to the person interacting with it based on his/her attributes. The devices are hence made smarter due to the enhanced computing power (in terms of processing and storage power) provided by the cloud. The architecture of Future Internet (FI) provided in section 2 provides a detailed explanation on the definition provided above.

A. Existing Methods and Our Contribution

Several access control schemes have been proposed such as Dynamic broadcast encryption technique [8] and role based access control (RBAC) [12] have addressed various issues namely traceability via trusted third party (TTP) and anonymity. Attribute based encryption [13, 14] has provided an effective mechanism to address the scalability and anonymity issues that has been adopted as a basis in [10, 11]. In addition many mechanisms have been proposed to enhance security mechanism against attacks such as replay attacks by leveraging the usage of timestamps [9]. However, the existing methods work upon the assumption that the cloud is honest (trusted) but curious, which is not an ideal case as the cloud operated by the cloud service provider (CSP) may lie outside the trusted domain of the concerned organization, thereby exposing the access policy to the cloud in order to make decisions upon providing access to requested users which paves a possible way for collusion attack (cloud colludes with an attacker and relaxes the policy providing illegal access to attacker).

Hence this paper adopts the attribute based encryption technique as a basis for the proposal. Further this method is enhanced by providing a means of hiding the access policy from the cloud while still leveraging the high computing power (in terms of processing and storage) of the cloud. In our method we expose the access structure only to the cloud that is sufficient for the cloud to perform its designated task and hide the access policy from the cloud. Further, user privacy is incorporated by allowing a user controlled environment where user decides upon how his/her attributes are to be exposed or portrayed for authentication during access control procedures (explained in section 2.2, Identity Portrayal). Finally a proposal about the idea of the TTP (trusted third party) that has the identity of the participating entities

(CSP and devices) registered with it that helps is establishing a mutual trust and traceability of activities during situations of misuse/fraudulence. Replay attacks are addressed by the usage of timestamps and DoS attack is to be taken care of by establishing a TTP (trusted third party) that helps to track down the activities of the cloud and also by using the session establishment technique.

B. Organization of the Paper

This paper has been organized as follows. In section 2 a basic architecture of the FI (Future Internet) which supports the definition of FI provided in section 1. Section 3 introduces us to the basics of Identity and access management and provides an explanation about Identity portrayal which is the basic idea behind the mechanism that is followed in our paper that allows users to control the exposure of his/her attributes during the access control process. Section 4 provides a high level depiction of where access control finds its place in FI that is explained in detail with a sequence diagram. Section 5 discusses in brief the existing works and also provides an evaluation of the work. Section 6 discusses the various attacks and threats that were considered during the evaluation of related work and also provide possible mitigation techniques that can be followed to address these issues. Section 7 finally proposes an architecture that addresses our major concern of collusion attack by incorporating the idea of separation access policy and access structure and also enhancing user privacy by providing a user controlled environment where user controls the portrayal of his attributes as explained in the scenario provided under section 3.2.

2 Architecture of Future Internet

This section provides one perspective about the organization of future internet. It also provides an insight into the components that find their place in the Future Internet (FI) and their appropriate roles. The basic architecture of the future internet is shown in Figure 1. It has at its lowest layer (i.e., layer 1) the devices (every possible real time object) such as smart phones, smart TV's etc. along with their unique identities (RFID, IPV6, biometrics etc.) to be able to stay connected and communicate to the network. The devices also have their connected sensors that help in the reception of stimuli from external physical environment.

The next layer (i.e., layer 2) is the access network layer that consists of access points including 3G, Wi-Fi and sensory networks (such as WSN gateways for the sensors) that helps to connect the devices over the internet.

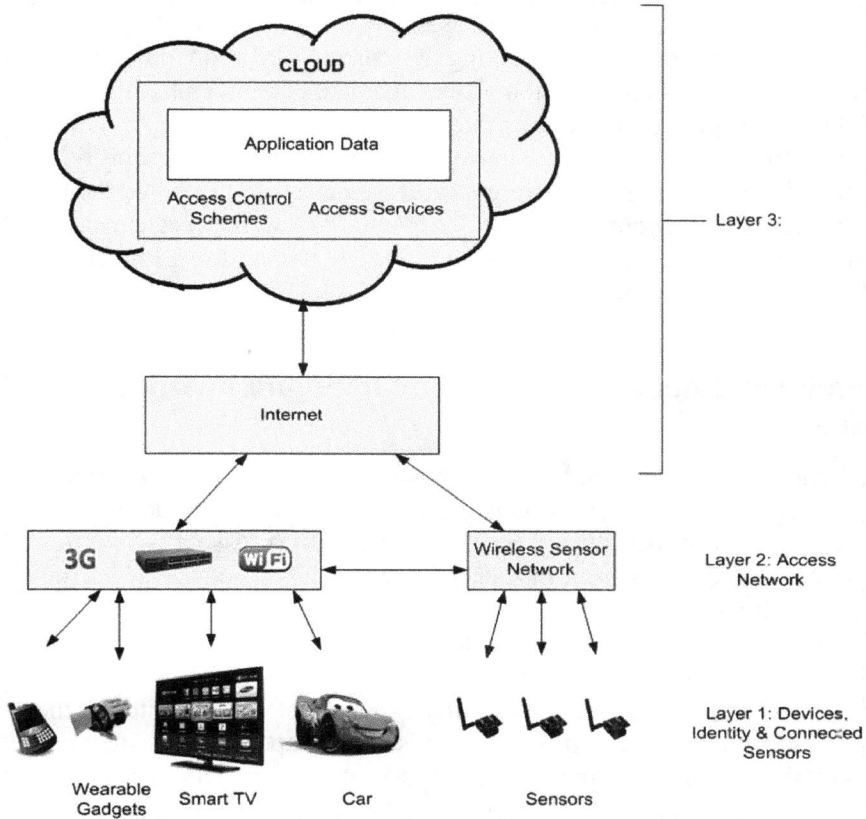

Figure 1 Architecture of future internet

The next layer (i.e., layer 3) consists of the internet that serves as a communication layer over which the cloud operates. The cloud is responsible for storage of data that are collected by the devices and associated sensors. The data for decision making related to access to specific applications based upon the stored (and received information from sensors) information also resides in the cloud. All the huge power consuming computations relating access management are moved on to the cloud which helps arrive at efficient decisions concerned with access control.

The cloud could be a private cloud (corporate wide) to store organizational private data or a public cloud that stores data that is mostly in the encrypted format and access control service data. The cloud here is looked upon as storage as a service and also in most cases serves as a medium to perform tasks that are heavily resource consuming. It is also looked upon to follow the access restriction rules and function as management authority that regulates access to the sensitive data stored by the customers.

Hence, from the above discussion we see that the key responsibility of protecting the data by regulating access (access control) to the stored data is borne by the cloud thereby leading to the point wherein the mechanism of regulating access (access policy) has to be implanted with a strong foundation thereby enhancing the protection mechanism.

3 Identity and Access Management in Future internet (IAMFI)

Identity and access management plays a major role in establishing security in FI. This section serves as an introductory explanation to the reader making him understand what Identity and access management is. Also a sub-section for Identity portrayal has been introduced because in section 7 an idea has been proposed wherein, during the phase when a request to access data on the cloud is sent to the cloud via a device that has the user attributes captured in it, an additional level of user privacy and security is provided where the user is given the flexibility to manage the exposure of his attributes to the cloud. For example, this user privacy establishes its importance in situations where certain social networking sites and email service providers try to access personal information in order to customize advertisements and contents for the appropriate users, which might be offensive for a few users who value their privacy [3].

3.1 Identity Management

Declaring the existence of a user to a system probably in the form of logon ID or username is Identification. Since user IDs are unique and not shared amongst users identification makes it possible to hold users accountable for the actions on the system. The users IDs are subjected to various policies (Eg. ID to meet minimum number of characters. IDs not to reflect usernames etc.) in order to enhance security.

Authentication is a verification done on the users' claimed identity and is implemented via passwords upon logon. Authentication falls under the following three types:

- Type 1: Something you know (e.g., personal ID number)
- Type 2: Something you have (e.g., ATM card)
- Type 3: Something you are (e.g., fingerprints)

Now, with Future Internet when every possible thing or object that is going to be connected to the Internet, identifying these items/object uniquely has to be done in an efficient and effective manner. The IPV6 addressing mechanism and other related RFID identifiers could be possible ways of providing unique ID's to the objects thereby taking care of the exhaustion of ID's available for identification currently in the IPV4 addressing. The existence of the devices and their associated sensors can be identified by means of the identification techniques such as IPV6, RFID, barcode and biometrics to identify and authenticate human beings. Also, Host Identity Protocol (HIP) is another method for assigning identities that includes security as an inherent part of the design and it separates host's identity from the topological location which provides a key basis for mobility of devices in future internet. It provides as a means of addressing the DoS (Denial Of Service) attack to a certain extent. Tools for managing the identities basically run as applications on dedicated servers or on the cloud. These interconnected objects are going to continually collect data, analyse them and use it in various decision making and management activities. Examples of things include people, all electronic devices (e.g. mobiles, smart TV's, food and even clothing).

3.2 Identity Portrayal

Identity management in FI not only concerns the assignment of unique identities to objects and managing their use in internet services, but also includes the management of presentation or portrayal of these identities online for communication. This portrayal of identity influences the enhancement of security and privacy to a greater extent. Management of Identity information by oneself helps to mitigate the threats posed to the access of data by the attackers [5].

The transformation of internet from an academic place to a public place of communication including e-commerce which involves business to business communication and also end users (customers) choices has led the world wide web to evolve into a place that allows exchange of information for everybody. Every user has his/her own profile that serves the purpose of identification. This profile is then used to portray the identity of users during information exchange or communication.

This self-portrayal of Identity follows that, in a communication session between two parties, the owner of identity or profile information can take control of exposing selected or all parts of his/her identity attributes for use in verification or processing by the other party. The attributes are then transmitted from the owner's site to the other party in a standard format. This selected and controlled revealing of identity attributes enhances privacy by preventing unnecessary or undesired processing of personal information.

To explain this, a simple but interesting real life example is shown in Figure 2 which helps us understand the self-portrayal of Identity and its associated privacy. In the figure, George, is a person from a village who is a Goldsmith by profession. He owns a jewellery and gift shop at the town where he sells his handmade products. In this scenario, George, his name is the core Identity by which he is universally identified. He has several other attributes that serves as an identity for him. However, George decides upon which attributes to expose to each of his different environments. For example, George is identified as George son of William along with his family details and personal home address by the people of his village (Persona 1). To the people of the town where John runs his business he does not expose his personal information but is identified with the attributes such as George the goldsmith, the quality of jewellery and gifts he sells, location of his shop and the shop name. In this way George keeps the exposure of his attributes by which he is identified under his control.

3.3 Access Control

Access control is the selective restriction based on certain rules. Access control helps to minimize access risk and fraudulent access to information Identification and authentication are the building blocks of access control systems.

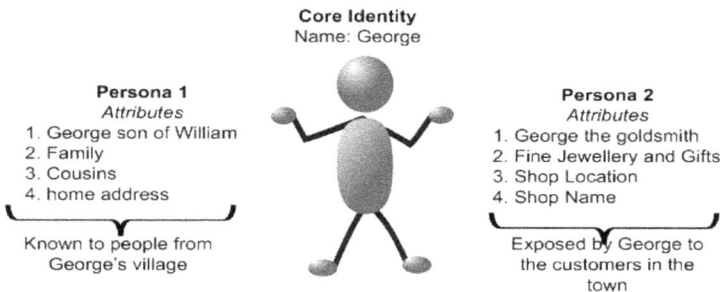

Core Identity
Name: George

Persona 1
Attributes
1. George son of William
2. Family
3. Cousins
4. home address

Known to people from George's village

Persona 2
Attributes
1. George the goldsmith
2. Fine Jewellery and Gifts
3. Shop Location
4. Shop Name

Exposed by George to the customers in the town

Figure 2 Identity portrayal

Access control belong to identity management by its very nature. Access control is required to preserve the integrity, confidentiality and availability of data over cloud (Future internet). Access control is implemented based upon the policies (high-level statements of management intent) that regulate the access of data and personnel who are authorized to receive the data.

There are a few things that must be considered when imparting access control mechanisms into systems such as

- Threats to the system (events that cause harm to the data)
- The system's vulnerability to the threats (lack of security mechanisms against threats) and
- The risk that might be materialized by the threats.

In the Future Internet, access control has to be implemented as a strong foundation at various interacting layers. At the core of the identity management for FI is the access control, wherein policies are defined that describe which devices and users are allowed to access data on the network and what tasks can be accomplished by the user. Users/customers can set up services and manage their hardware through the cloud access control. Threats coming from networks such as spoofing data should be safeguarded by using access control techniques.

Access control services and schemes (predefined rules and strategies set by data and application owners) are called by the applications and related stored data to decide upon authentication and provisioning of services to requested users who satisfy/dissatisfy the specified strategies.

3.4 Current Trends and Standardizations

In the Future Internet, apart from software based in IAM, the trend is moving towards biometrics and hardware re-inforced security. For example we already have digitally signed BIOS in the newer chipsets [6]. Another approach is going away from passwords entirely and moving the identity and access management in to the devices as passwords tends to be weak and people can't remember passwords for multiple sites [7]. Several attempts in developing an open standard for Identity Management is under work by organizations such as the Cloud Computing work group- Cloud/SOA-Security and architectures for Identity Management. Many security and privacy enhancing mechanisms for ubiquitous cooperation and interaction are being dealt with currently. Many areas supporting the existence of Future Internet and security thereby have been identified and Cloud computing is one such area which has hit its mainstream by recent releases of vendors. The ETSI Specification group has been working in order to develop a common view on Identity and access management in Future Internet, in terms of protocols and architectures relating to network of connected devices. In the event of such attempts of unification of diverse network topologies and multiple devices the FI architecture developed must deal with the identity and access management issues handily. The CSA projects funded under FP7 has jointly ventured into supporting the Standardization activities within the Future Internet Assembly (FIA) that covers the various domains including the security (including access management), internet of things and cloud computing.

4 A Vision of Access Control in Future Internet (FI)

This section provides a glimpse of the importance and place held by access control in FI. This section shows how the identity and access control mentioned in section 3 fits into the real life scenario of FI, where we have shown numerous devices and facilities (in Figure 3 and Figure 4) that are uniquely identified by assigned identities as objects. Access control explained in section 3 is found to have its place in FI where a user is provided restricted access (access control) to various facilities as in Figure 3 and also in Figure 4 where user controls the operations various devices (interaction between devices) within the house. These devices and facilities shown in Figure 3 and Figure 4 form the lowest layer (layer 1) depicted in Figure 1 under Section 2 (basic architecture

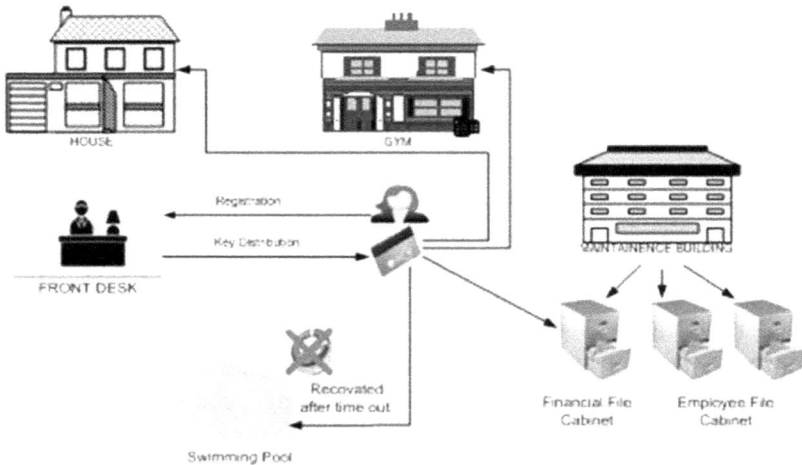

Figure 3   Motivation - building maintenance

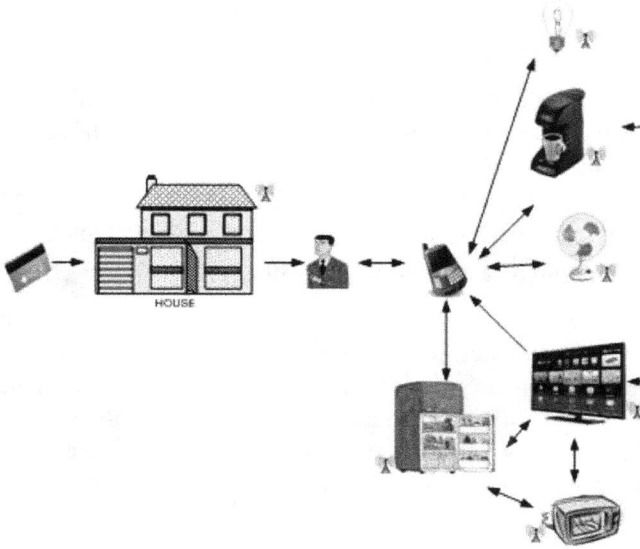

Figure 4 Motivation – device to device communication

of Future Internet). The motive of considering access control in FI (Future Internet) is explained in the following paragraph.

We are moving away from the traditional method of access control i.e., using usernames and passwords as passwords are weak and it is cumbersome to remember due to the transformation of communication methods between entities. Internet of Things is an aspect of Future Internet where numerous devices and sensors will be interacting with each other and the surrounding environment. Just by knowing the scale of this FI aspect the traditional methods cannot prove to handle the access and security issues effectively, thereby motivating us towards working upon the evolution of an access control mechanism that can be applied to the FI scale.

4.1 User Centric Vision

In a buildings' maintenance and operation scenario, there could be a lot of facilities like the swimming pool, library, the maintenance building with houses a lot of sensitive filling cabinets (e.g. financial, corporate, etc.) and a bunch of different homes/apartments. This scenario is depicted in Figure 3 wherein a visualization of one of the possible ways of fitting access control to regulate access to resources is portrayed.

The access and security issues that we have to address may include similar aspects as follows:

- If somebody is employed over the pool for a couple of months (say only for summer) then we need to make sure that they do not have access to the pool after their employment ends.
- Any user who has been given access to the maintenance building then it is very certain that we must take care to see that people are not accessing the information that they are not entitled to.

In order to address these issues we could get rid of the all the keys and provide a key card to access all the facilities and buildings.

Doing so we could ensure that at the end of a user's employment with the pool the key card would automatically revoke access to the pool. This is basically done based upon a role or attributes of the users, meaning that if any user has been given access to the maintenance building and if they were part of the financial group/role then she would be able to access the financial filling cabinet while the others who do not satisfy the role/attribute would be denied access.

The access cards provide facility to keep track of all of the different facilities that users can access at any given point of time. It is also possible to report/track on the different activities (access to facilities). Further, the users contact a help desk in order to request for keys in order to be able to access the facilities across the community. It is also possible to track the workflow involved in granting the access.

4.2 Extended Device to Device Communication – Vision of Future Internet Scenario

Apart from user centred access. Device to device communication is another phenomenon that is addressed in the FI. This device to device communication which forms the essence of Future Internet and consequently invoking a strong need for access control that serves the purpose of addressing security issues relating to Future Internet (FI) is shown in Figure 4 as an extension to the scenario in Figure 3.

Moving on to elaborate upon the above scenario, taking into consideration the possible activities that could take place inside the house, once the user has identified himself as the authorized user he gains access to use the facilities inside the house and also control them.

For example, in Figure 4, the user can turn on/off lights and other electronic equipment via the identity established through his smart phone. A fridge can further communicate to the smart TV that serves as a display for the user to view the contents present in the fridge. The fridge can also communicate with the smart TV to display the recipes that could be cooked with the available grocery and also display a manual for the recipes suggested. The TV in turn could communicate to the microwave oven by providing the oven with the manual for the recipe that has been chosen by the user and start the cooking as specified by the manual. The fan with the associated sensor for the temperature of the room can sense the environment and work accordingly.

A visitor would be allowed to access the house and its facilities by sending him a one-time access code that they could use to get initial access to the door (opening door) by operating it over their smart phones. Other facilities such as access to light up specific rooms, access the microwave oven and TV could be granted upon approval by the current owners/occupants of the house.

4.3 Sequence Diagram Showing the Flow of Actions for Access Control in Distributed Environment

The general sequence of actions that takes place during access control in shown in Figure 5. Initially the user registers himself with a key distribution centre and obtains keys/Unique ID. Here user and devices have been separated because in ideal cases there are situations when the user needs to first identify himself as the authorised users to gain access to the devices and perform further operations using them.

The user then sends his identification information to the device as a request to gain access to the device. The devices have with them associated ID's that identify them uniquely as with that of the users and also has associated sensors that help to sense and gather information related to the environment or other user related information/activities.

The devices upon receiving the access request from the user might not have the capability to verify the validity of the user and hence sends the received information over to the cloud where the user related information and access policy that governs the approval of access is stored. The cloud then verifies the details received and based upon the validation sends a reply to the device whether to permit access or deny access. If the user has been authenticated as a valid user by the cloud the device in turn authenticates and allows access to the user. The user then uses the device to perform further actions.

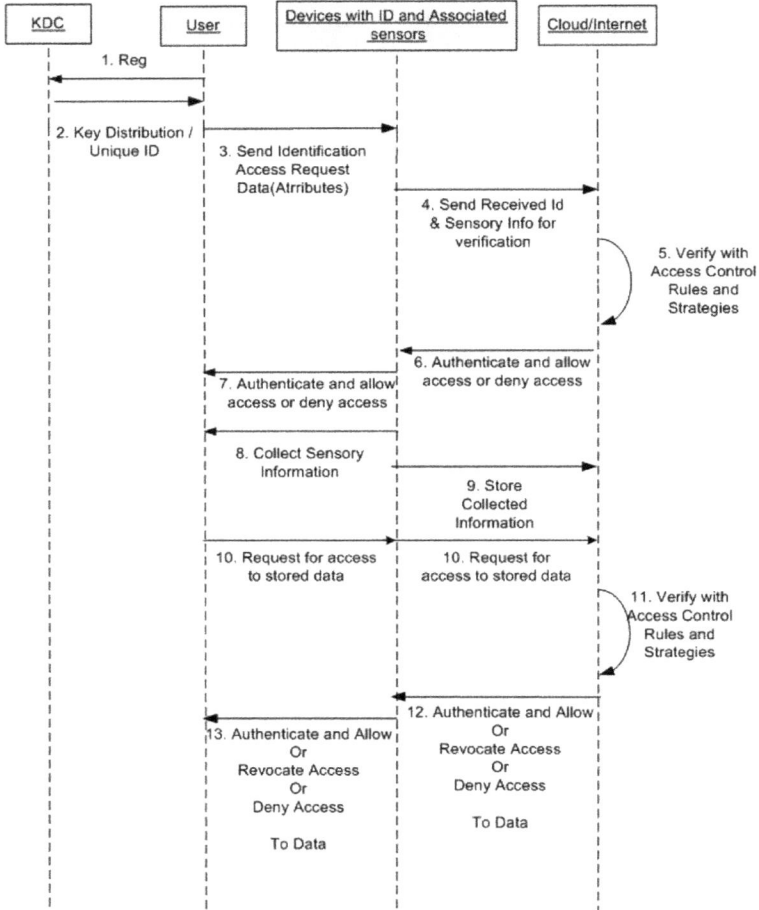

Figure 5   Sequence diagram showing the flow of actions for action control in distributed environment

The user then uses the device to collect information and store the information on the cloud. He also specifies the access policy that tells the cloud about the users who are authorised to access the data based upon which the cloud accepts or rejects the request to access from users. When a second user requests for access to the stored data on the cloud by providing his identity information with the request, the identity is verified by the cloud against the access policy that stands guard for the stored data, and access is provided to those users whose attributes match the access policy or denied access otherwise.

5 Related Works and Evaluation

Several access control mechanisms have been introduced in order to provide effective means of providing access to stored data over the cloud. In [8], the author has proposed an efficient solution for supporting multiple authorized users to read the outsourced data as well as modify the data over the cloud. The author has established the mechanism of maintaining the anonymity of the users who share the data over the cloud by leveraging group signature and dynamic broadcast encryption technique. Adding and revoking users has also been addressed efficiently. However, the technique has adopted a centralized approach to distribution of keys which solely depends upon the group manager which is not an ideal case. Also the existence of a predefined group might not always be possible when cloud computing is under consideration. In [9], the author has adopted the Attribute based signature and attribute based encryption technique, wherein anonymity of the data owner from other users has been preserved, however the traceability has also been provided by the trusted third party under the situations of forged/illegal activity. It also addresses the situation where not only the owner of the data makes changes to the data over cloud but also the authorized users are allowed to update the data and upload it to the cloud. The technique has adopted a decentralized approach towards key distribution.

In [10], the author has adopted the Key rotation and broadcast encryption technique. Due to the existence of a trusted third party between the cloud and the organization there is an establishment of mutual trust between the two parties, which addresses the risks of collusion and forgery from both sides. Also, the trusted third party provides traceability of any illegal activities. The broadcast encryption technique has made the scalability aspect efficient. However, anonymity of users/data owners which is required in certain situations has not been considered in this technique. The technique has also not addressed the possibility of replay attacks that would support forged activity. The user revocation is partial wherein there is trade-off between the computational cost of re-encryption (minimized-lazy revocation) and the availability of data for revoked users i.e. the revoked users can gain access only to the stale data (not updated versions). The mechanism has guided the possibility in which a single owner updates the data and uploads it to cloud while other users can only read the data upon decryption. Multiple authorized users who would find a need to update the data over the cloud is not addressed.

In [11], the author has adopted the ACV-BGKM (Access control vector broadcast group key management) technique which follows the idea of sharing some secrets to users based on the identity attributes they have and later allow them to derive actual symmetric keys based on their secrets and some public information. Thus the adding/revoking users or updating access policies could be performed efficiently. However, the provision to trace any malicious activity by the cloud or the users has not been addressed. In [12], the author has adopted the Role Based Access control Model (RBAC) model and the role based encryption cryptographic technique. The introduction of a separate private cloud to hold sensitive information and usage of public cloud only to share data to users outside the organization has provided a means to prevent external collusion attacks. Also, the organization's sensitive information (eg. the role structure etc.) is hidden from the distrusted public cloud. However, the mechanism poses a difficulty to scalability due to a possibility of role explosion. Also, the mechanism has taken a centralized approach wherein a user has to register to the organizations authority to obtain keys to access and decrypt the required data. The mechanism has guided the possibility in which a single owner updates the data and uploads it to cloud while other users can only read the data upon decryption.

In [13], the author has proposed a new multi-authority CP-ABE (cipher text policy attribute based encryption) technique to efficiently handle adding and revoking users in an environment that needs to support rights to all authorized users to read as well as update his/her part of data over the cloud. It has adopted a decentralized approach towards key management. However, anonymity of users/data owners which is required in certain situations has not been considered in this technique. In [14], A Hierarchical Attribute-Based Solution for Flexible and Scalable Access Control in cloud Computing, the author has proposed a hierarchical CP-ABE by adopting the traditional CP-ABE thereby inheriting its security schemes enhanced with a hierarchical set of users in order to overcome the limitation of flexibility in implementing complex access control policies. The access control technique has realised a scalable, flexible and fine-grained access control over the data stored on cloud. It has also dealt with user revocation efficiently by employing multiple value assignments for access expiration time. However situations at which a user/owner of the shared data would not like his/her identity to be revealed to other users in the group have not been considered.

All of the above techniques except [12] have not considered the limitation wherein the organizational sensitive information (access policy and attributes) are exposed to the cloud. However in [10], the technique emphasizes that the organization has a private cloud to hold sensitive data and a public cloud to store data, which does not solve the sole purpose of relieving the organization from taking care of availability and maintenance of resources. Another limitation is that all the above techniques have not considered the situations when a CSP could actually refrain from providing services to the authorized users, deletion of data for accommodating space etc. (DoS and mutual trust). It has also not considered the situation where there might be certain situations when authenticated users might also want to update the data and move the copy over the cloud. Table 1 and Table 2 summarize the above discussion and give a comparison between the various key attributes that are to be considered while performing access control to stored data on the cloud. In Table 1 provides a comparison between the major attributes that come into picture when access control is considered in cloud and Table 2 compares the attributes related to key management such as encryption computation cost for the keys and data generated as well as the provision of updating data to users (number of users allowed to read and write) and also the revocation of users and its associated feasibility in key management. In Table 2 the abbreviations in the last column are M-multiple, R-read, W-write and 1-Single.

6 Attack Modelling

This section provides a short description of the attacks such as DoS (Denial of Service), Collusion and Replay attacks. It provides an insight into the possible threats that might cause the system to experience the associated attacks. A modelling of the attacks is depicted via sequence diagrams and mitigation techniques that address such attacks are also explained. This modelling provides a good understanding towards the threats posed to a system and the security measures to be taken when expecting their resultant impact as attacks to the system.

6.1 DoS (Denial of Service) Attack

Denial of service is an attack on a cloud service attempting to indefinitely or temporarily stop the cloud services to the intended or authorized users. Cloud ignoring or losing the updated data by users and hiding loss of data can also be considered as DoS attack, as the cloud is in fact not providing the required service to the user.

Table 1   Comparison chart of related work major attributes

Paper No	Attack Resistance			Anonimity	Traceability	Mutual Trust	Fine Grained	Scalability	key Distribution	Confidentiality of Access Policy and Attributes
	DOS	Replay	Collusion							
5	No	Not efficient	Yes, but collusion between cloud+revoked users and data+authorized users agains cloud not addressed	Yes	Yes-Group manger	No cloud accountability	Yes	More efficient	Centralized	Access policy and attributes known to cloud
6	No	Efficient	Yes, but collusion between cloud+revoked users and data+authorized users agains cloud not addressed	Yes	Yes-trusted third party(only for users not cloud)	No cloud accountability	Yes	Yes, but incurs more computational cost	Decentralized	Access policy and attributes known to cloud
7	Yes	No	Yes (due to trusted third party traceablity ob both sides)	No	Yes-trusted third party	Accountability from both sides	Yes	More efficient	Centralized	Access policy and attributes known to cloud

#										
8	No	Yes	Yes, but collusion between cloud+revoked users and data+authorized users agains cloud not addressed	Yes	Not addressed	No cloud account-ability	Yes	More efficient	Decentralized (keys computed at time of decryption)	Access policy and attributes known to cloud
9	Not addressed	No	Yes (cloud cannot collude with revocated users)	No	Yes	No cloud account-ability	No	Not efficient due to possibility of role explosion	Centralized	Organisation's sensitive structure information (rolcs etc..) are hidden fromthe cloud
10	No	No	Yes, but collusion between cloud+revoked users and data+authorized users agains cloud not addressed	No	Yes	No cloud account-ability	Yes	Yes	Decentralized	Access policy and attributes known to cloud
11	No	Yes (expiration time for small window of vulneribility)	Not addressed	No	Yes	No cloud account-ability	Yes	More Efficient	Centralized	Access policy and attributes known to cloud

Table 2   Comparison chart of related work (key management attributes)

Paper No	Encryption Computation Cost	User Revocation	R/W Access
5	Independent of no of revoked users	Yes, but requires multiple encryted copies of same data file	M-R-M-W
6	Most expensive operations moved to cloud	Yes, but requires multiple encryted copies of same data file	M-R-M-W
7	Cost of reencrytion minimized (lazy revocation)	Partial-Lazy revocation trade off	1-W-M-R
8	Minimized	Yes	1-W-M-R
9	Minimized	Yes, but requires multiple encryted copies of same data file	1-W-M-R
10	Minimized computational cost as compared to the traditional CP-ABE (ciphertext attribute based encryption)	Yes by attribute authority	M-R-M-W
11	Minimized when compare to traditional ABE	Yes, but requires multiple encryted copies of same data file	1-W-M-R

6.1.1 Threats leading to denial of service

- Improper Design Techniques of Cloud. Where the cloud doesn't foresee the resource required for the user. In case of less storage space the cloud either deletes the contents or does not allow the updates.
- Typically an external attacker would use SQL Injection, Regular Expression Parsing, Buffer Overflow, Unexpected Exceptions and excessively large file uploading by attacker.

Overwriting of old data(knowingly/unknowingly) with new data(probably data belonging to another customer) can be considered as an indirect deletion of old data. Thereby, leading to the denial of servicing request for old data that resided at the cloud prior to overwriting. Hence, there is always a possibility of cloud deleting data knowingly or unknowingly and it is therefore a valid assumption.

On August 8, 2008 a 45% of customers lost their data from an online storage company Linkup and it was shutdown. There was finger pointing between it and another company Nirvanix on which Linkup had relied upon [15]. Secondly forensics on cloud is not yet evolved to imply cloud legally that

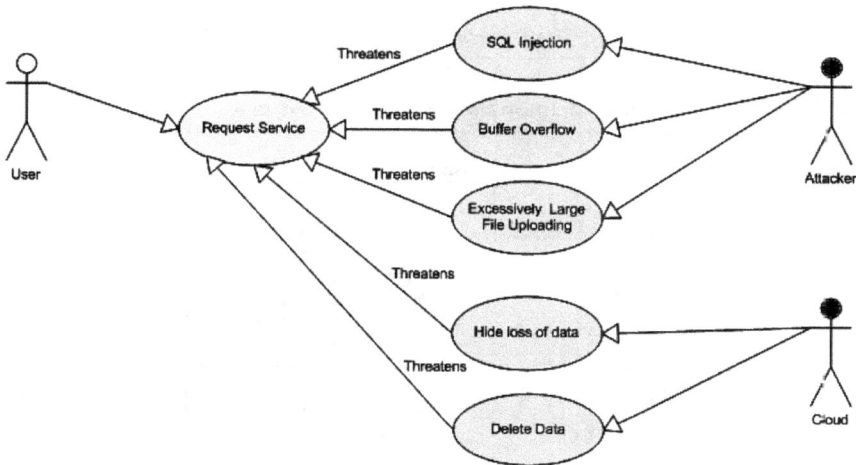

Figure 6   Mis-use case diagram depicting dos attack

cloud has accidentally or due to inefficient infrastructure design had caused data loss by removing, overwriting existing data. [16].

The Mis-Use Case depicting a typical scenario where an attacker uses the cloud system vulnerabilities to deny a legitimate user access to the cloud is shown in Figure 6. Here, a legitimate user requests for a service from the Cloud, meanwhile an attacker performs undesirable activities such as SQL injection, buffer overflow, or exclusively large file uploading that bombards the communication channel of the CSP (Cloud service provider) and thereby poses a threat to the responsive action of the CSP in servicing the request posed by the legitimate user. Also, activities such as hiding any loss of data or deliberate deletion of data by the cloud, in the former case to maintain reputation and in the latter case to make space for more customers poses a threat to the request of data posed by the user. In either case the cloud refrains from providing the requested service to the legitimate user which is denial of service.

6.1.2 DoS (denial of sevice) attack modelling

The sequence of actions depicting the DoS attack is shown in Figure 7 and Figure 8. In Figure 7, the owner/device sends his relevant information to the cloud as proof of identity and upon verification is given authorized access to store encrypted data along with the access policy on the cloud. When a second legitimate and authorized user/device requests access to the stored data, the

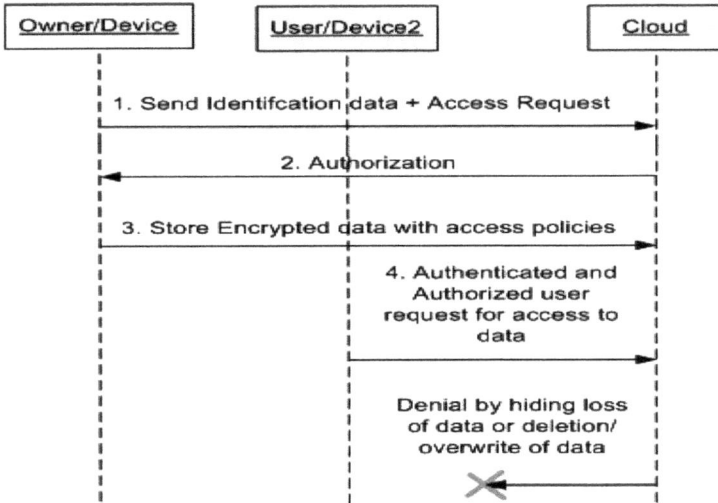

Figure 7   Sequence diagram depicting DoS by cloud

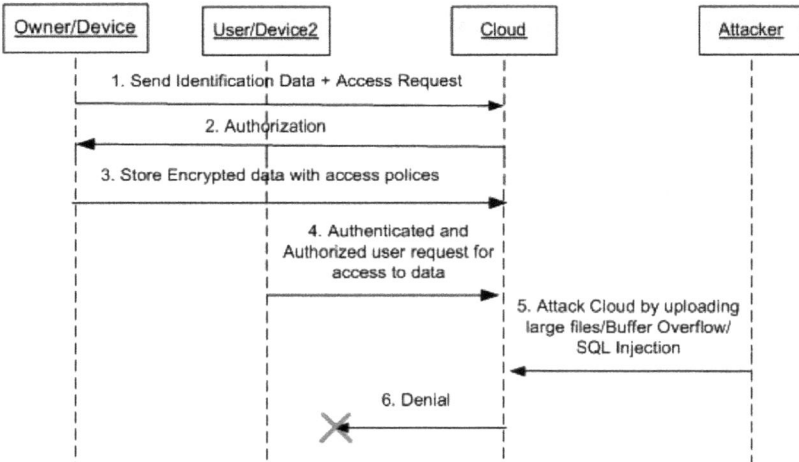

Figure 8   Sequence diagram depicting DoS by attacker

cloud refuses to provide the service. The reason for the denial could be the result of unacceptable activities such as loss of data due to certain internal reasons or deliberate deletion of data by the cloud. Additionally the cloud might as well claim that the data never resided with him.

In Figure 8, the owner/device sends his relevant information to the cloud as proof of identity and upon verification is given authorized access to store encrypted data along with the access policy on the cloud. When a second user/device is a legitimate and authorized user requests access to the stored data the cloud refuses to provide the service due to unacceptable activities performed by the attacker such as SQL injection, buffer overflow and uploading of large files.

6.1.3 DoS mitigation techniques:

- TTP (Trusted Third Party) to Monitor the activities of participants (Cloud and the users)
- Installation of monitoring s/w tools that identifies DoS patterns and takes steps accordingly.
- Cloud Design has to be validated by TTP for reliable storage and operational failures so that hiding loss of data and updates to data can be monitored. The Cloud has to be built with secured software interfaces & APIs so that the attacker doesn't take advantage of the cloud's vulnerabilities and stun the system functionality.

6.2 Collusion Attack

In Collusion Attack, two or more participants join hands in order to develop privileges that satisfy the required access policies that stand guard for the protected data. In the mis-use case diagram below both cloud and the malicious user collude to threaten user's access policy and authorization info.

6.2.1 Threats leading to collusion attack:

- Malicious Insider and untrusted cloud: Malicious Insider can join hands with cloud in order to gain access over the sensitive information. The cloud having known the access policy and being trusted to provide access to users relaxes the policy to allow access to malicious insiders with insufficient authentication information.
- External Attackers and untrusted cloud: Malicious external users gain unauthorized access to user's account information and manipulate the data. The Cloud Service Provider partners with this user knowingly or unknowingly to sabotage confidential information.

The mis-use case diagram in Figure 9, shows the possible activities of the malicious user and its collusion with the cloud to compromise on access rules

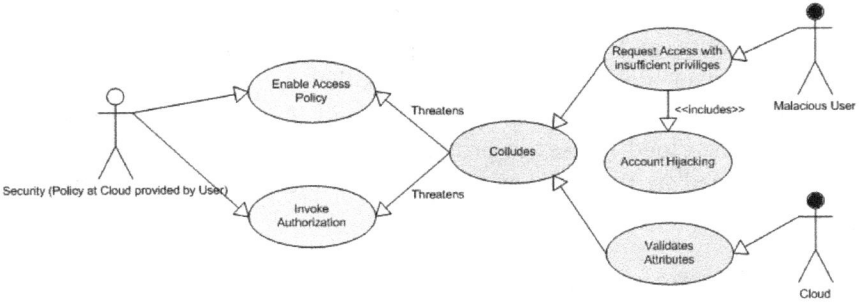

Figure 9 Mis-use case diagram depicting collusion

that threatens the normal security activities that executes in order to safeguard the data. Here, the malicious user with insufficient or totally unmatched privileges requests access to a confidential data stored in the cloud. The cloud on the other hand who is responsible to validate users based upon the access policy and provide access to only the authorized users colludes with the malicious user and relaxes the access policy restrictions allowing the malicious user to portray himself as a legitimate user (mask himself under a valid account-account hijacking) and provides illegal access to the data.

6.2.2 Collusion attack modelling

The sequence of actions depicting the Collusion attack is sown in Figure 10. Here, the owner/device sends his relevant information to the cloud as proof of identity and upon verification is given authorized access to store encrypted data along with the access policy on the cloud. When an attacker/malicious user (insider or an outsider) requests access with insufficient privileges the cloud relaxes the access policy constraints and deliberately colludes with the malicious user/attacker and allows him to access the data as if he were a legitimate user.

6.2.3 Collusion attack mitigation technique

Trusted Third Party can be used to audit the participants' (cloud and the users) activities for fraudulent behaviour, so as to increase the transparency and to trace the source of fraudulence with the help of the identities of the participants that are registered with the TTP.

Nothing is safe in the cloud because "Cloud is assumed to be honest "which is not an ideal case [8, 9]. On this ground a method wherein we could hide even the access policy from the cloud and thus prevent the facility of cloud

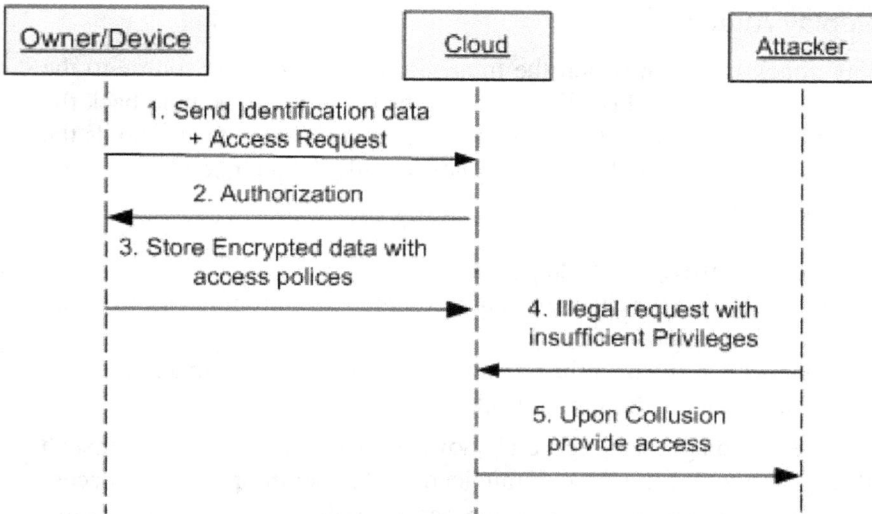

Figure 10 Sequence diagram depicting collusion

allowing unauthorized access to stored data which is the crux of the proposed idea in Section 7 has been proposed.

And as mentioned in the paper in Section 6.2.1, collusion attack is described as one happening when situations wherein cloud joins hands with malicious insiders and also the external attackers, and even though the cloud knows that the user is not an authorised one to access data, there are possibilities wherein the cloud could relax the access policy and provide access. This happens because the cloud is not a trusted one all the time. We cannot work upon the assumption that the cloud is honest but curious as is the case with most of the related work discussed in the paper. Relaxation of access policy occurs because the access policy is by default open to the cloud and the cloud is the one that is trusted by the data owner to take care of regulating access to the data and in situations where the cloud is outside the trusted domain of the owner the data is always under threat. Hence an idea has been proposed wherein we make use of cloud to only store data and do some complex resource consuming operations while the decision about regulating access still lies in the hands of the owner or personnel within his trusted domain, by providing only the access structure to the cloud and hiding the access policy. The entire working is explained in Section 7.

6.3 Replay Attack

A replay attack is one in which the transmitted data from the source to the destination over a channel is sniffed by an attacker wherein he hold back the data midway for a stipulated amount of time, tampers the data and sends the tampered or inappropriate data that was not actually the intended information to the destination.

6.3.1 Threat leading to replay attack

- An attacker sniffs the channel for any confidential communication or user related information.
- An attacker waiting over the transmission channel tries to manipulate the transmitted data by holding it back for some time.

The mis-use case diagram in Figure 11 shows the activities of the attacker such as sniffing and monitoring the communication channel in order to get access to confidential data during transmission and alteration of data that threatens the security and confidentiality of the encrypted data that was intended to be sent to the destination by the source.

6.3.2 Replay attack modelling

The sequence of actions depicting the Replay attack is shown in Figure 12. Here, the owner/device sends his relevant information to the cloud as proof of identity and upon verification is given authorized access to store encrypted data along with the access policy on the cloud. A second legitimate and authorized user/device initiates a request for access to the data stored on the cloud and the cloud upon verification of the authenticity initiates the transmission of requested data over the communication link. An attacker who has been monitoring the link holds back the transmitted data tampers or changes the entire information and sends the tampered/inappropriate data to the legitimate user/device which could

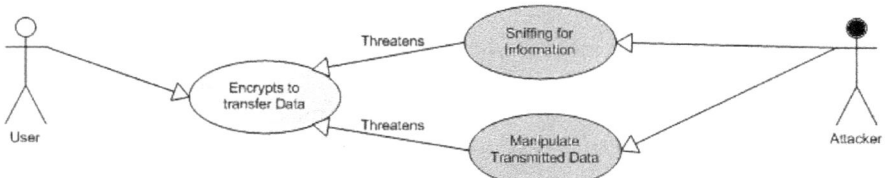

Figure 11   Mis-use case diagram depicting replay attack

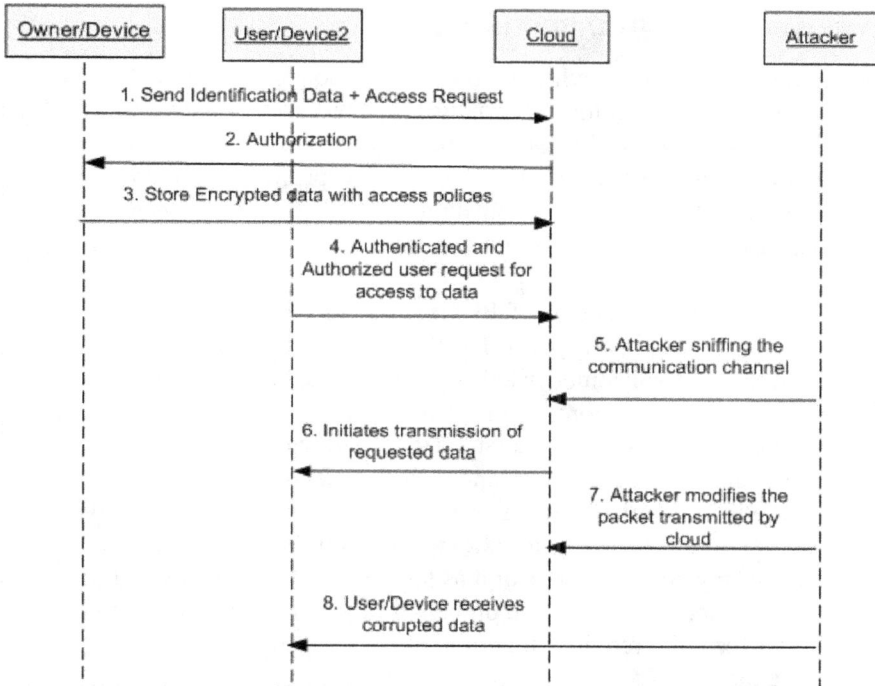

Figure 12 Sequence diagram depicting replay attack

hamper normal functionality. Alternatively this tampering of data over the communication link can also happen when the legitimate user/device sends some data to be stored over the cloud so that when some other user wants to access the data that was originally sent by the owner, is given inappropriate data.

6.3.3 Replay attack mitigation technique

When the message is sent from the user to the cloud, it should include a time stamp or other similar techniques that would pose control based on a predefined (minimal) duration of time that is set at the time of transmission from sender. This is an estimation of the time required for information to reach the destination during normal operation. If this time is exceeded the information is assumed to be tampered and hence it is retransmitted upon suitable request from destination or is ignored at the destination.

7 Proposed IAMFI Architecture

IAMFI addresses the issue wherein the access policy and the attributes (sensitive organizational information) are exposed to the cloud. IAMFI attempts to provide a mechanism of privacy of the attribute information by incorporating the idea of separation of profile information and profile structure [18–20] into the attribute based encryption method for providing access to users who request access to the stored data on the cloud.

As mentioned in architecture of FI given in Section 2, we see that the sensitive accessible data along with the access policy lies in the cloud. So the point of access control comes in existence when the devices communicate with the cloud to access the stored data. Also from the paper [21] and from the discussions from Section 4.2 we understand that the devices serve as a medium of communication between the users and the cloud. These devices connected over the internet along with cloud makes the future internet. Devices could be autonomous as in the case where a light switch controls light in a room over IP. The light lets any device to control it as long as the device is located in the same room, or devices could operate on behalf of users by capturing the user attributes to perform the required function.

Therefore, the proposed IAMFI has considered the later scenario, wherein the device captures the user attributes and serves and medium of access to the cloud. In the Figure 13, we have represented the IoT (Internet of Things) assuming that the devices are connected over the internet and unique Identities are assigned using various techniques such as RFID, Near Field Communication, IPv6, HIP etc. The following scenario explains the high level architecture of IAMFI.

The proposed IAMFI architecture is shown in Figure 13 wherein, the owner (organization) of the data has several attribute authorities (AA) depending upon the domain under which the various categories of users who access his data. For example, an organization can have users who belong to the same organization (owner is the AA for his personal domain i.e. Personal AA in Figure 13), the researchers who require the organizational data9i.e, Research AA in Figure 13) and common people who are indirectly related to the organization (i.e., Public AA in Figure 13). In all such cases the appropriate attribute authority maintains the management of the users who fall under his category or domain.

The AAs (personal and other attribute authorities) distribute keys to the users who wish to obtain access to the data. The attribute authorities (AA) are also responsible for making decisions upon whether or not to provide access to a user/device that requests access to stored data.

The users who wish read and write access first register themselves with the attribute authority (AA) and obtain appropriate attributes that determine his/her access to data. The AA is responsible for secret key distribution so that the overhead of assigning keys to all the users collectively does not bother the owner (Personal AA).

The Owners (Personal AA) encrypt their data under a certain selected set of attributes. The owner also specifies an access structure tied with the encrypted data that is stored on to the cloud. We consider the attributes of users to be associated with certain keys [18]. For Example, the pairs that say [University = XUY, Location = SC Position = Professor]. Here the values [Position, University and Location] are taken as the keys and the values associated with them [XUY and SC, Manager] are the actual values/attributes/data

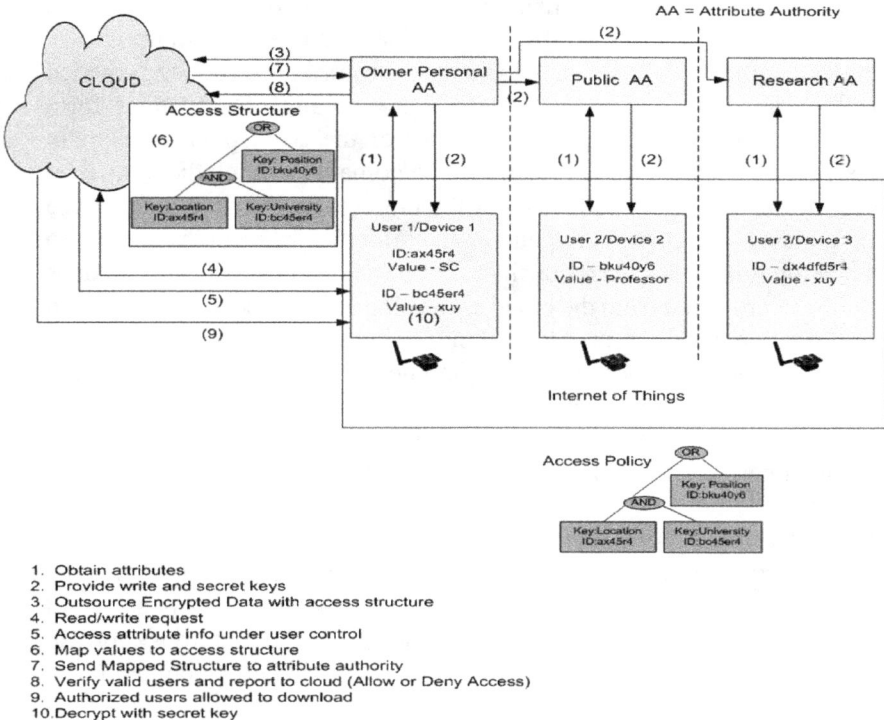

1. Obtain attributes
2. Provide write and secret keys
3. Outsource Encrypted Data with access structure
4. Read/write request
5. Access attribute info under user control
6. Map values to access structure
7. Send Mapped Structure to attribute authority
8. Verify valid users and report to cloud (Allow or Deny Access)
9. Authorized users allowed to download
10. Decrypt with secret key

Figure 13 IAMFI architecture

that are required for providing access. To enhance privacy we only provide the access structure to the cloud as shown in Figure 13. For example, [location and university or position].We associate IDs with the keys that reside in the access structure, for example, [key = Location ID = ax45r4] and also associate same IDs to the corresponding data/value/attribute that reside at the user's associated devices for example, [Value = SC, ID= ax45r4]. It is to be noted here that the sensitive data that is the exact attributes or roles that are intrinsic to the organization are not exposed to the cloud.

When a user requests for data for read/write access, the cloud will request for attribute verification. A session is established where the user is in full control while the cloud accesses the attributes of the user. The cloud checks for an ID match to retrieve the values/attributes to the keys that are mentioned in the attribute structure. For example cloud uses the ID [key=Location ID=ax45r4] that is mentioned in the access structure to compare it against the IDs present at the user's devices. Upon finding [Value = SC, ID= ax45r4] that has a matching ID the cloud maps the value/attribute to the key Location in the access structure thus forming the <key,value> pair i.e., <Location,SC>. The cloud uses these attributes to map them to the attribute structure provided by the Owner and computes a Boolean value or any other similar form of representation and sends it to the Owner or the corresponding domain AA(attribute authorities) for verification. The domain to which the computed value has to be sent is made known to the cloud whenever the user/device sends a request to the cloud by appending the information about the domain to which it belongs to. This data is considered to be a non-sensitive data that does not need to be protected from the cloud as the cloud cannot pose a threat based upon the location information. Also, all the attribute authorities lie within the trusted domain of the owner of the data and hence the owner also shares the access policy against which decisions are to be made for allowing access, with the various individual AAs. Once the data has been sent the session is closed and any data corresponding to the attributes of the user is deleted from the cloud storage.

The owner or the corresponding AA checks the received computed value against the access policy that it has at its end and if a match is found then the owner sends a reply to the cloud to allow access to the user or deny access in case there is no match found. In this way the cloud is prevented from knowing the access policy or attributes that satisfy the specified access policy. Once access has been granted to the user/device, s/he can download the data and decrypt it using the secret keys received from the attribute authority. It is to be noted here that the identity assignment to devices has not been included in the scope of this paper. It is assumed that devices that exist online already have been assigned unique Identities.

Hence, the proposed IAMFI architecture helps to avoid the possibility of collusion attack as mentioned in Section 6.2 by hiding the exact access policy from the cloud. Hence the cloud has no possible chance of learning the exact access policy that provides access to stored data, additionally since the decision about providing access to data still lies in the hands of the owner or other trusted personnel within his trusted domain the possibility of cloud colluding with the malicious attackers and providing unauthorised access is being taken care of. Further replay attacks are to be taken care of by appending a timestamp T and a MAC (message authentication code) to the request/data that is sent. Also a predefined threshold value is provided. If the intended destination finds that the timestamp T is greater than the predefined threshold it is assumed to be held back midway and tampered and is discarded. Finally the DoS attack is to be taken care of by establishing a TTP (trusted third party) that helps to track down the activities of the cloud. Further the DoS attack that happens due to the attack on a cloud service by devices attempting to indefinitely or temporarily stop the cloud services to the intended or authorized users is taken care of by the session establishment that is proposed in this section and the associated device identities. It is easy to manage the access using one ID per session and also address the access of the same ID to the same resource by regulating it to one session at a time.

8 Conclusion and Future Work

Thus IAMFI provides a mechanism of privacy of the attribute information by incorporating the idea of separation of profile information and profile structure into the attribute based encryption method for providing access to users who request access to the stored data on the cloud. IAMFI also relieves the owner from the overhead of managing the user registration and key management activities. The attribute authority will be able to trace the activity of the user in case of malpractice with the registered Identity of the user. It is also possible for the personal attribute authority to trace the activity of the cloud service provider by allowing the attribute authority to have the ID of the CSP registered with it. Alternatively, a trusted third party (TTP) can also be introduced wherein the users and the CSP register their IDs with the TTP. In the event of any fraudulent activity the TTP serves as a mediator who would track the IDs and identify the person responsible for the fraudulence.

In future extension to this work will be done by developing appropriate algorithms and models for implementing IAMFI with suitable key management techniques that takes care of user attributes/identities and appropriate key generation in our following papers. We will also be extending our work by developing detailed protocols to address the various attacks that are mentioned in this paper will be developed.

9 List of Abbreviations

Acronym	Expansion
IAMFI	Identity and Access Management in Future Internet
FI	Future Internet
IoT	Internet of Things
CSP	Cloud Service Provider
AA	Attribute Authority
ACV-BGKM	Access control vector broadcast group key management
RBAC	Role Based Access control Model
CP-ABE	cipher text policy attribute based encryption
DoS	Denial of Service
TTP	Trusted Third Party
KDC	Key Distribution Center

References

[1] http://en.wikipedia.org/wiki/Internet
[2] http://www.nets-fia.net/
[3] http://www.washingtonpost.com/blogs/the-switch/wp/2013/11/04/how-we-know-the-nsa-had-accessto-internal-google-and-yahoo-cloud- data/
[4] http://www.csoonline.com/article/205053/the-abcs-of-identity-management
[5] https://vsis-www.informatik.uni-hamburg.de/getDoc.php/publications/201/BaierKunze04-INetSec.pdf
[6] http://www.zdnet.com/access-control-changes-a-must-for-future-safe-internet-vint-cerfsays[7000018569/
[7] HolgerKinkelin, HeikoNiedermayer, Ralph Holz, and Georg Carle, 'TPM-based Access Control for the Future Internet', Network Architectures and Services TechnischeUniversitätMünchen

[8] Xuefeng Liu, Yuqing Zhang, Member, IEEE, Boyang Wang, and Jingbo Yan, 'Mona: Secure Multi- Owner Data Sharing for Dynamic Groups in the Cloud', IEEE Transactions On Parallel and Distributed Systems, Vol. 24, No. 6, June 2013.

[9] SushmitaRuj, Member, IEEE, Milos Stojmenovic, Member, IEEE, and AmiyaNayak, Senior Member, IEEE, 'Decentralized Access Control with Anonymous Authentication of Data Stored in Clouds', IEEE Transactions On Parallel And Distributed Systems, Vol. 25, No. 2, February 2014.

[10] AyadBarsoum and Anwar Hasan, Senior Member, IEEE, 'Enabling Dynamic data and indirect mutual trust for cloud computing storage systems', IEEE Transactions OnParallel And Distributed Systems, Vol. 24, No. 12, December 2013.

[11] Mohamed Nabeel, Member, IEEE, Ning Shang, and Elisa Bertino, Fellow, IEEE 'Privacy Preserving Policy-Based Content Sharing in Public Clouds', IEEE Transactions On Knowledge and Data Engineering Vol. 25, No. 11, November 2013.

[12] Lan Zhou, Vijay Varadharajan, and Michael Hitchens, 'Achieving Secure Role-Based Access Control on Encrypted Data in Cloud Storage', IEEE Transactions On Information Forensics and Security, Vol. 8, No. 12, December 2013.

[13] Kan Yang, Associate Member, IEEE, XiaohuaJia, Fellow, IEEE, KuiRen, Senior Member, IEEE, Bo Zhang, Member, IEEE, and RuitaoXie, Student Member, IEEE, 'DAC-MACS: Effective Data Access Control for Multiauthority Cloud Storage Systems', IEEE Transactions On Information Forensics And Security, Vol. 8, No. 11, November 2013.

[14] Zhiguo Wan, Jun'e Liu, and Robert H. Deng, Senior Member, IEEE, 'HASBE: A Hierarchical Attribute-Based Solution for Flexible and Scalable Access Control in Cloud Computing', IEEE Transactions On Information Forensics and Security, VOL. 7, NO. 2, APRIL 2012.

[15] BRODKIN, J. Loss of customer d.ata spurs closure of online storage service 'The Linkup'. Network World (August 2008).

[16] CLOIDIFIN.http://community.zdnet.co.uk/blog/0,1000000567,2000625 196b,00.htm?new_comment

[17] Mervat Adib Bamiah, Advanced Informatics School Universiti Teknolog, Malaysia Kuala Lumpur, Sarfraz Nawaz Brohi Advanced Informatics School Universiti Teknologi, Malaysia, Kuala Lumpur, 'Seven Deadly Threats and Vulnerabilities in Cloud Computing', International Journal Of Advanced Engineering Sciences and Technilogies, Vol No. 9, Issue No. 1, 087 – 090, 2011

[18] BjoernWuest, Olaf Drogehorn, KausDavid, 'Architecture for profile translation', Supported in part by European Union Information Society Technology, February 04 2005.

[19] R.M. Arlien, B. Jai, M. Jakobsson, F. Monrose, M.C. Reiter, 'Privacy–preserving global customization', In Proceedings of the second ACM conference on Electronic commerce, Minneapolis, *USA*, p.176–184, October 2000.

[20] S.Riche, GBrener, M.Gittler, 'Client-side ProifleSorage: a means to put user in control', Public Technical Report. Heweltt Packard Laboratories Grenoble, November 2001.

[21] Jan Janak, Hyunwoo Nam, and Henning Schulzrinne Columbia University, 'On Access Control in the Internet of Things', February 15, 2012.

Biographies

Nancy Ambritta P. graduated in Computer Science and Engineering from Anna University, Tamil Nadu, India in the year 2010. She is currently pursuing Masters in Computer Engineering at Smt. Kashibai Navale College of Engineering, Pune. Her research interests are cloud security and the Future Internet.

Poonam Railkar received the Masters in Computer Networks from Pune University Maharashtra, India in the year 2013. She is currently working as an Assistant Professor at Smt. Kashibai Navale College of Engineering, Pune. She has published 6 journals and 3 conference papers. Her research interests are Mobile Computing and Security.

Parikshit N. Mahalle received PhD (Wireless Communication) from CTIF, Aalborg university, Aalborg, Denmark and is IEEE member, ACM member, Life member ISTE and graduated in Computer Engineering from Amravati University, Maharashtra, India in 2000 and received Master in Computer Engineering from Pune University in 2007. From 2000 to 2005, was working as lecturer in Vishwakarma Institute of technology, Pune, India. From August 2005, he is working as Professor and Head in Department of Computer Engineering, STES's Smt. Kashibai Navale College of Engineering, and Pune, India. He published 40 papers at national and international level. He has authored 5 books on subjects like Data Structures, Theory of Computations and Programming Languages. He is also the recipient of "Best Faculty Award" by STES and Cognizant Technologies Solutions. His research interests are Algorithms, IoT, Identity Management and Security.

Traffic Offload Guideline and Required Year of the 50% Traffic Offloading

Shozo Komaki, Naoki Ohshima and Hassan Keshavartz

Malaysia-Japan International Institute of Technology, Universiti Teknologi Malaysia, komaki@ic.utm.my, naoki.osm@ic.utm.my, hassan@mjiit.com

Received: November 22, 2013; Accepted: June 1, 2014
Publication: July, 2014

Abstract

Smart Phone and tablet terminals are widely accepted into mobile society and support wireless cloud service effectively. Terminals generally adopt flat rate tariff and the traffic is increasing rapidly. To solve this problem, new technology developments and new spectrum resource allocations and assignments are intensively executed. However in quite near future, traffic will overcome this action. This paper proposes the traffic offloading to microcells and give numerical guideline of offloading ratio that minimize the total radio base station cost under the existing spectrum resource allotment. First, offload guideline is derived based on the Japanese congested area case study in Shibuya ward, and this guideline is translated and generalized to global circumstances. Using the guideline, the required offloading year is calculated for the high population cities or wards in the world. From the results of the analyses, it is shown that the traffic offloading to microcell is necessary in near future. This result is valuable and inevitable to minimize increasing spectrum allotment to the existing mobile service. To monitor the offload ratio, it is better to analyze social bigdata and the carrier's bigdata. In the final part, example of Draft Question for ITU-R is proposed.

Keywords: Traffic offloading to microcell, Offload guideline, Radio base station cost minimization, 3GPP TS LIPA SIPTO.

Journal of ICT, Vol. 2_1, 37–64.
doi: 10.13052/jicts2245-800X.212

1 Introduction

Wireless broadband access is rapidly penetrating into mobile communication service and terminals, such as smart phone, tablet PC,wireless cloud and so on. Rapid growth of wireless broadband access generates very heavy traffic on the mobile networks. It is well known that the bottleneck for heavy traffic is comes from radio spectrum limitation. To handle heavy traffic, new technology and new radio frequency band allocation have been developed intensively. They were powerful motivation of the global research and development on wireless technologies, including signal processing and control.

However the traffic growth of smart phone has different nature from the existing mobile service, such as voice and mail service. They are similar to fixed broadband service, in that the flat rate tariff is introduced to enhance user utility for various cloud services. For the user utility and mobile carrier benefit, new fruitful applications should be supported by wireless cloud, as same as in fixed networks. Such a increasing wireless broadband traffic will overcome existing technology development and radio spectrum allocation in near future.

Proper solution to solve this bottleneck, it is requested to implement new microcell radio base statin (RBS) to the congested area. This is called as the traffic offloading to microcell. [1–3] 3GPP published TS 23.829, TS 23.859 and TS 22.220, in that H (e)NB, Local IP Access (LIPA) and Selected IP Traffic Offload (SIPTO) are described. H(e)NB includes Home Node B (HNB) and Home e-Node B (HeNB), they are working on 3GPP frequency bands and set in the home or the office building. SIPTO is a method where portions of the IP traffic on a H(e)NB access or cellular network is offloaded to a local network, in order to reduce the traffic load on the existing 3GPP networks, called as core system or macrocell system. In the TS 23.402, the non-3GPP access, such as trusted/non-trusted wireless LAN (WLAN) access, is also defined. In the following part of this document, the SIPTO using H(e)NB and the trusted/non-trusted WLAN access are called as femtocell system and WLAN access system, respectively. Also it is defined that microcell system includes femtocell and WLAN access. Existing large area system operating in 3GPP band is called as macrocell.

Technological issues have already been prepared as is mentioned, however the traffic offloading to femtocell is not widely implemented, by the reasons of the spectrum coordination between macrocell and femtocell, the difficulty of air interface update on femtocell RBS and the difficulty on RBS management and maintenance issue. Considering the existing WLAN, one of promising traffic offloading may be trusted/non-trusted WLAN access. Any way, it is necessary to enhance and promote the offloading to microcell including femtocell and WLAN, by setting up the offload guideline to congested cities. Obtained result suggests that the congested city in the world requires early offloading in the near future.

In this paper, traffic offloading to microcell is analyzed, and the cost minimized offloading guideline is proposed. Analyses are done under the condition of total RBS cost minimization and existing spectrum resource is fixed and limited. The optimum offload guideline is calculated by using real congested city parameter in Shibuya-ward, Tokyo Metropolitan area, Japan. Obtained results are generalized and applied to the congested cities and wards in the world.

The guideline calculates required 50%-traffic offloading year for various cities and wards.

Definition of required 50%-offloading year means that the communication quality of service (QoS), such as the latency time and the throughput in the data communication, is seriously degraded, if the offloading is not put into force. In this analyses, all the data is assumed to be the Packet Switched (PS) based data traffic, and the throughput decrease and the latency time increase in the PS mean that the voice call blocking rate increases in the circuit switched (CS) system, when PS and CS co-share bandwidth of trunk line. So if traffic offloading to microcell is not taken after the required 50%-traffic offloading year, the communication QoS, such as latency time and call blocking, seriously decreases not only in data communications but also in voice communications.

If there is no traffic offloading guideline, mobile carriers select another business tactics. To solve the bottleneck in easygoing manner, it may be considered to introduce the non flat-rate tariff or the limited monthly quota. This strategy is not good selection for wireless cloud era, from the viewpoint of the competition among mobile carriers and also social utility. Another solution may be traffic control and/or shaping. The concept of network neutrality already set up in fixed networks, and it requires fairness, reasonable and transparent control. Non flat-rate tariff implementation to users may not be easy issue under the competition among several mobile carriers.

If the new macrocell RBSs are additionally build in the congested area to maintain QoS, the RBS invest cost without offloading becomes higher comparing with the traffic offloading to microcell, and results in the lose competition power among carriers.

Chapter 2 mentions current and forecast of estimated monthly traffic per capitain Japan. Chapter 3 describes the structure of traffic offloading.

Chapter 4 describes analysis model and principle to minimize total cost of mobile base stations. In this chapter, Clark's city model and cost ratio of macrocell and microcell base station is introduced to generalize the analysis. Chapter 5 shows the result of Shibuya ward, and the generalized value of traffic offload guideline is proposed. Chapter 6 derives the required 50%-traffic offloading year, based on the proposed generalized guideline. And some examples of the required offloading year are shown for various high population density cities and wards in the world, and it is shown that the time limit for offload is critical and very near in the congested cities.

2 Traffic Growth of Mobile Service

Japanese mobile traffic is monitored and prospected in the various organizations [4–6]. Figure 1 shows the monthly data size per capita (GB/month).

The Ministry of Internal Affaire and Communications Japan (MIC), summarized the existing network traffic report in 2004. In those days, compact HTML based on the i-Mode service and text mail were dominant, and the

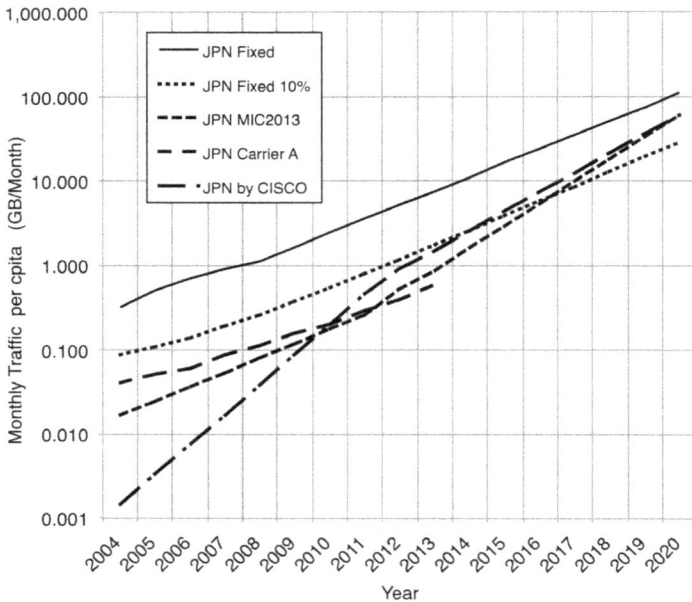

Figure 1   Japnese mobile traffic status and prospect

traffic is smaller comparing with the today's traffic. In 2013, the real traffic is monitored and estimated by MIC. Also two mobile carriers in Japan prospected from their observed traffic in 2007. Those are shown in the figure, and lies little bit smaller than 10% of the fixed internet services. Traffic of fixed broadband service increases annual rate of 46%. From the estimations, target traffic of 3years later from now, i.e. 2017, reaches 10GBytes/month. Traffic estimated by Cisco is also illustrated in the figure. The annual growth rate (ARG) is as same as that measured by Japanese organizations. As shown in the figure, traffic growth is very rapid and the monthly traffic volume reaches to near volume of the fixed internet services. This is the motivation of the guideline setup.

3 Traffic Offload

Conceptual schema of traffic offload is shown in Figure 2. Figure 2(a) shows existing macrocell system without traffic offload, and all users utilize same macrocell mobile base station. Figure 2 (b) shows traffic offload to microcell system. In high population and/or high traffic area, microcell base stations, for example trusted/untrusted WLAN access is implemented, and traffic is offloaded to microcell mobile station. Usually, microcell base station is connected through fixed broadband networks and vertical handover between microcell and macrocell mobile networks are essential. Microcell base stations are implemented in high traffic area in highpriority. As the result of the priority, the traffic offload ratio is larger than microcell area ratio, as is shown in Figure 4.

4 Analysis Model

Evaluation parameter for offload guideline is the total RBS cost minimization under the limitation of existing radio spectrum. Outline of the model and the used parameters are summarized in Table 3. Detailed issue of the analysis model is shown in Section 4.1 through 4.4.

4.1 City and Population Model

Shibuya ward, Tokyo metropolitan in Japan is selected as a typical high traffic area. The obtained result is translated and generalized to the other global area shown in the chapter 5. Profile of Shibuya ward is shown in Table 1. Offered communication traffic is estimated from daytime population, because the peak traffic at go home hour is closely relating with a day time population.

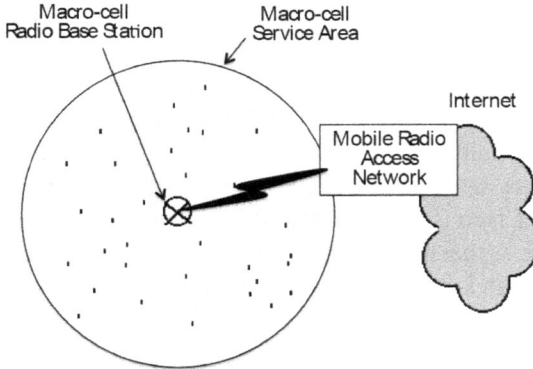

Figure 2   (a) Existing macrocell system

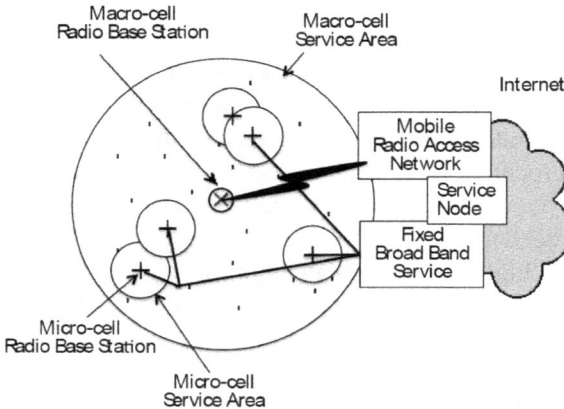

Figure 2   (b) Traffic offloading to microcell

Real population distribution of Shibuya is like Figure 3(a). Some shopping mall and vicinity of railway stations show traffic peaks at several points and peak concentration ratios are different among them. This is generalized to equivalent Clark's city model [7] shown in Figure 3(b). This city model has circler area that has same space, same population of Shibuya ward and modified equivalent population concentration ratio at the city center.

Using Clark's Model, population density $f(r)$ is exponential and given by,

$$f(r) = \eta e^{-dr} \tag{1}$$

Table 1 Profile of shibuya ward in Tokyo metropolitan, Japan

Daytime Population N_d [person]	559,000
Area S_{area} [m²]	15,110,000
Mobile Subscription rate R_m [%]	78
(National Mean Value)	
Equivalent City Radius of Shibuya	2,193
(Circular Model) R [m]	
Specified Carrier Share R_d [%]	53.7
(Tokyo area)	
Number of Users P_{area}	234,157
Population Concentlation Ratio	3
(Clark City Model) β	
Time Concentlation Ratio β_t	3
City parameter η [person/m²]	0.1395
City parameter d	1.8477×10^{-3}

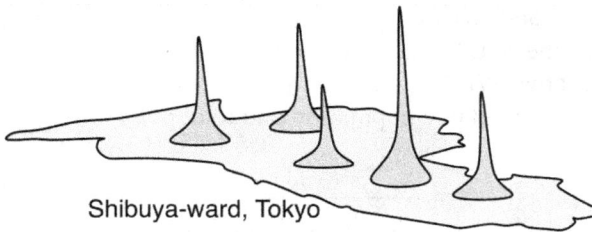

Shibuya-ward, Tokyo

Figure 3 (a) Real population distribution of shibuya ward, Tokyo metropolitan, Japan

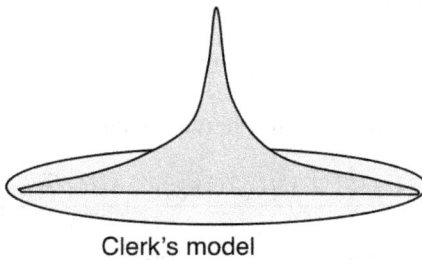

Clerk's model

Figure 3 (b) Equivalent population model of clark model for calculation

where, r denotes distance from city center, and η, d are parameters, those are calculated as follow.

If the total daytime population N_d exists in the all city area S_{area}, and city radius is $R = \sqrt{S_{area}/\pi}$, and introduce the population concentration ratio β defined by the ratio of maximum population density at the city center and average population density, then peak population density η at the city center is shown

$$\eta = \beta \frac{N_d}{S_{area}} = \beta \frac{N_d}{\pi R^2} \qquad (2)$$

Population N_d exists in the city area S_{area}, then

$$N_d = \int_S f \, dS = \int_0^R 2\pi r \, f \, (r) dr = \frac{2\pi\eta}{d^2}(1 - (1 + dR) \, e^{-dR}) \qquad (3)$$

Substitute (3) into (2) then,

$$(dR)^2 + 2\beta(1 + dR)\, e^{-dR} - 2\beta = 0 \qquad (4)$$

From the equation, parameter d is determined by using population concentration ratio β and city radius $r = \sqrt{S/\pi}$.

Supposing that the WLAN access point is installed first priority at the highest population area, WLAN available population ratio δ, that is defined by the ratio of WLAN available population and total population, is

$$\delta = \frac{\int_0^{r_{BS}} 2\pi r f(r) \, dr}{\int_0^R 2\pi r f(r) \, dr} \qquad (5)$$

where, r_{BS} denotes equivalent radius of WLAN equipped circle area. If the WLAN installed area ratio α is defined by the ratio of WLAN installed area and total city area, and then

$$\alpha = \frac{\pi r_{BS}^2}{\pi R^2} \qquad (6)$$

Using Equation (5) and (6), relation between δ and α is given as follow.

$$\delta(\alpha) = \frac{1 - (1 + dR\sqrt{\alpha})\, exp\,(-dR\sqrt{\alpha})}{1 - (1 + dR)\, exp\,(-dR)} \qquad (7)$$

Figure 4 shows the WLAN access available population ratio $\delta(\alpha)$, for-Shibuya ward, where $S = 15,110,000 \, [m^2]$, and population concentration ratio $\beta = 1, \ 3, \ 6, \ 9$.

As is seen in the figure, in case of population concentration ratio of $\beta = 3$, 30% of population, i.e. traffic, can be offloaded, by installing WLAN into the 20% of total area. In case of more population concentrated case of $\beta = 9$, the 60% of traffic can be offloaded to WLAN by installing it into only 20% of total area.

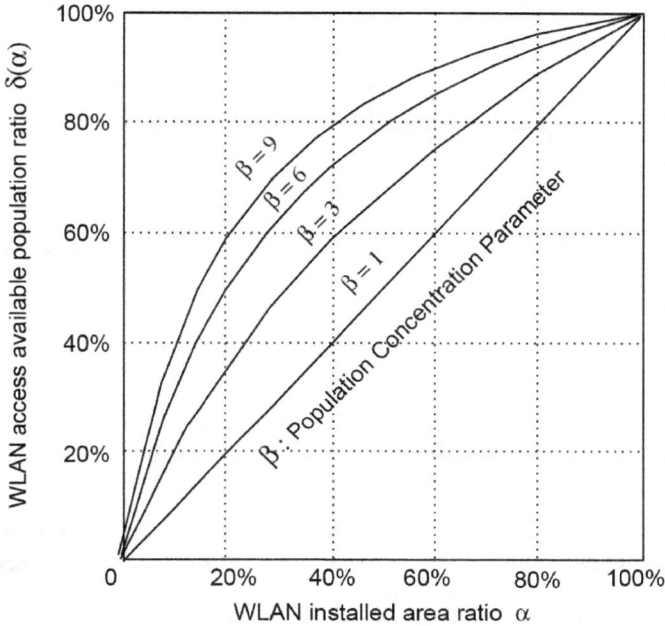

Figure 4 Relation between WLAN installed area ratio α [%] and WLAN available population ratio δ [%]

4.2 Radio Base Station Model

Radio base station model and specifications are summarized in Table 2. Macrocell system utilizes LTE and WLAN utilizes IEEE802.11n operated in 5GHz to avoid interference from existing 2.4GHz users. Effective throughput is estimated by 20% of maximum throughput.

4.3 Traffic Model and Latency Time

Offered traffic of macrocell and microcell, D_{macro}, and D_{micro}, respectively, are

$$D_{macro} = (1 - \delta\,(\alpha))\frac{A\ P_{area}}{n_{macro}} \tag{8}$$

$$D_{micro} = \delta\,(\alpha)\,\frac{A\ P_{area}}{n_{micro}} \tag{9}$$

Table 2   Radio base station model for analysis

	Macro-cell		Micro-cell
System Name	LTE		IEEE802.11n
Bandwidth [MHz]	10*		20***
Number of RF Carriers [ch]	2**		3
Number of Sectors	3		1
MIMO	2x2		3x3
Maximum Capacity [Mbps]	86		156
Effective Throughput [%]		20	
[Mbps]	17.5		31
Effective System Capacity [Mbps]	35		93
Cell Radius [m]	500		20
Area Size [m^2]	785,398		1,256

*: one of pair band for frequency division duplex (FDD).
**: assigned channels to the one operator, i.e. equivalent to total.
***: time division duplex (TDD) bandwidth.

where, A, P_{area}, n_{macro} and n_{micro}, are the traffic from one user, the number of total users in the city for a specific mobile carrier, the number of macrocell RBS and the number of microcell RBS, respectively.

Required throughput of each macrocell and microcell RBS, ρ_{macro} and ρ_{micro} are

$$\rho_{macro} = (1 - \delta(\alpha)) \frac{AP_{area}}{n_{macro} T_{macro}} \tag{10}$$

$$\rho_{micro} = \delta(\alpha) \frac{AP_{area}}{n_{micro} T_{micro}}, \tag{11}$$

where T_{macro} and T_{micro} is the effective system capacity of each macro and microcell RBS, respectively.

To analyze the internet traffic, it may be better to use fractal or log-normal distribution not Poisson process, however, in this analysis, the widely used M/M/1/FCFS (Fast Come First Served) packet switched (PS) system is assumed, for the easiness of calculation and first order approximation. In this analyses, all the data is assumed to be the PS based traffic, and throughput decrease and latency time increase in PS means equivalent for the voice call blocking in circuit switched (CS) system, when PS and CS co-share bandwidth of trunk line.

Queue length $E[n_t]$ is

$$E[n_t] = \frac{\rho}{1 - \rho}, \tag{12}$$

where ρ is average throughput of the system. From Littele's Law,

$$E\,[n_t] = \lambda\,E[t], \tag{13}$$

where $E[t]$ and λ denote the mean latency time and the mean packet arrival, respectively. The mean packet size and the maximum capacity are denoted by s and T, then the mean through put ρ will be given by

$$\rho = \frac{\lambda\,s}{T} \tag{14}$$

Then the mean latency time $E[t]$ is given as follow.

$$E[t] = \frac{s}{T(1 - \rho)} \tag{15}$$

For the web browsing, packet size distribution is known as log-normal distribution, and observed mean value is 11.18, and variance is 2.24 [9]. Mean data size s is $s = exp\,(11.18+(2.24)^2/2) = 880$ [kB].

Using Equation (10), (11) and (15), macrocell and microcell latency time, $E[t]_{macro}$ and $E[t]_{micro}$ are as follows,

$$E[t]_{macro} = \frac{s}{T_{macro} - (1 - \delta\,(\alpha))\frac{AP_{area}}{n_{macro}}} \tag{16}$$

$$E[t]_{micro} = \frac{s}{T_{micro} - \delta\,(\alpha)\,\frac{AP_{area}}{n_{micro}}} \tag{17}$$

4.4 User Utility and Required Latency Time

Relation between user utility $U(t)$ and latency time t is measured and show exponential relation, and given by

$$U\,(t) = g\,e^{-ht} \tag{18}$$

where $g\ and\ h$ are parameters, and in case of fore ground picture browsing, following values are observed by using mean opinion score (MOS) experiment, based on ITU-R BT500-10 double-stimulus impairment scale (DSIS). User utility value U, i.e. MOS value, has in 0 (not allowable) to unity (no degradation), and allowable limit is normally set 0.6 (degraded but allowable). Figure 5 shows experimental results for various download case of Web Foreground (Web-F), FTP-Foreground (FTP-F) and FTP-Background (FTP-B). Web-F is normal user style for smart phone terminal, and most intolerant case. In this case, user utility U is shown as follow [8],

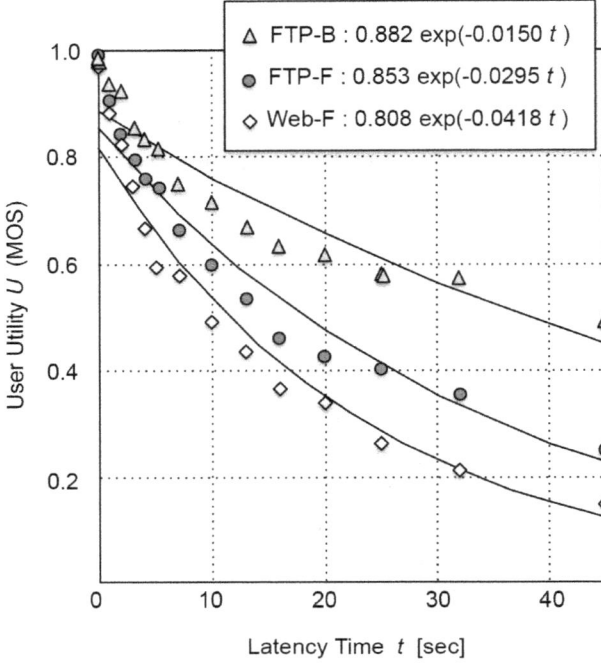

Figure 5   User utility and latency time

$$U\ (t) =\ 0.808\ exp(-0.0418\ t\ ) \tag{19}$$

i.e. $g = 0.808$ and $g = 0.0418$.

If we set required MOS value $u = 0.6$, then the objective mean latency time $E[t]$ is determined by the following Equation.

$$E[t] = -\ \frac{1}{h}\ ln\ (u/g)\ =\ 6.9\ [sec] \tag{20}$$

From the Equation (16) and (20),

$$E[t]_{macro} = \frac{s}{T_{macro} - (1 - \delta\,(\alpha))\frac{AP_{area}}{n_{macro}}} \leq\ -\frac{1}{h}ln\ (u/g) \tag{21}$$

Then necessary base stations n_{macro} is given as follow.

$$n_{macro} \geq\ \frac{(1 - \delta\,(\alpha))AP_{area}}{T_{macro} + s\,h/\ ln\ (u/g)} \tag{22}$$

Also the necessary base stations n_{micro} is given as follow.

$$n_{micro} \geq \frac{\delta(\alpha) AP_{area}}{T_{micro} + s\,h/\,ln\,(u/g)} \tag{23}$$

WLAN installed area ratio is determined by α, so the following condition should be satisfied.

$$\alpha \leq \frac{n_{micro}\,S_{micro}}{S_{area}}, \tag{24}$$

where S_{micro} denotes the area of each microcell.

4.5 Total Cost Minimization by Traffic Offloading to WLAN Access

Total cost minimizing method and parameters are summarized in Table 3.

Mobile base station selection strategy and cost minimization are as follows.

1. Daytime population density is estimated from the Clark's city model. User density of the specific mobile carrier is calculated using their share.
2. Area traffic is calculated from the average user traffic per capita and the user density. Busy hour traffic concentration ratio of $\beta_t = 3$ is considered in this stage, which is shown in Table 1.
3. Required latency time, i.e. 6.9 [sec], which is derived from the utility function shown in Equation (20), decides the required throughput and number of base stations for macrocell and microcell, i.e. Equation (22) and Equation (23), respectively.
4. Minimize the sum of base station costs, and then decide offload ratio.

To minimize total RBS cost, cost ratio of macrocell RBS and micocell RBS is important parameter. RBS cost ratio C is defined by

$$C = C_{macro}/C_{micro} \tag{25}$$

where, C_{macro} and C_{micro} are the cost of each macrocell RBS and the cost of each microcell RBS, respectively.

Total RBS cost, $Cost$ is given as follow,

$$
\begin{aligned}
Cost &= C_{macro} \times n_{macro} + C_{micro} \times n_{micro} \\
&= C_{macro}\frac{(1 - \delta(\alpha))AP_{area}}{T_{macro} + s\,h/\,ln\,(u/g)} + C_{micro}\frac{S_{area}}{S_{micro}}\alpha \tag{26}
\end{aligned}
$$

Table 3 Mobile base station selection model	

City Model and Population Density	Metroplotan Area (Shibuya ward, Tokyo, Japan)
	● Area Model: Clark Circular City Modle,
	● Area: Equivalent to Shibuya ward,
	  Radius 2.193 km, Area size 15.1 km^2
	● Daytime Population: 559,000 persons
	● Population Density:
	  Population decreasing exponentially toward city fringe
	  City center consentration factor = 3

Traffic Model	● Data Traffic
	  Quewing System (M/M/l /FCFS)
	● Poisson Distributed Traffic,
	  Mean traffic per user A is valiable parameter
	● Busy hour traffic ratio = 3

User Utility and Required Latency Time	● User utility function (Handa:2007):
	  Web foregrownd download
	  $U(t) = 0.808 \exp(-0.0418t)$
	  ITU-R BT. 500-10 (DSIS:Double-stimulus impairment scale)
	● Latency time: Less than 7 seconds, in that user utility U more than 0.6.

Radio Base Station Implementation Methodology (Offload Guideline)	Mobile Base Station Implementation Methodology
	● No spectrum allocation of new RF carriers and new bands
	● Implement Traffic Offload System to microcell base station
	● Number of Microcell and Macrocell Basestations
	  ● Cost minimum implementation of macro and micro cell BS
	  ● Fullfill user utility of more than 0.6
	  ● BS implementation cost ratio C is variable
	  C = Macrocell BS Cost / Microcell BS Cost

Mobile Service Operator's Profile and Base Station Equipment	Market share of one operator: 53.7% (Tokyo vicinity)
	Mobie Subscription ratio: 78% (Avarage of Japan)
	Radio Equipment:
	● Macrocell: LTE 10MHz, 2 pair bands, MIMO 2x2
	  Maximum speed: 85.7Mbps/band
	  Effective throughput: 20%
	  Area size (radius): 500 m
	● Microcell:IEEE 802.11 n, 20MHz, 4 bands
	  Maximum speed: 150 Mbps/band
	  Effective throughput: 20%
	  Area size (radius): 20 m

Then optimum WLAN installed area ratio α is derived as Equation (27),

$$
\alpha = \begin{cases} 0 \\ \frac{\pi}{d^2 S_{area}}\left(ln\left(\frac{\eta C S_{micro}}{T_{macro}+sh/\,ln\,(u/g)}\right)\right)^2 \\ 1 \end{cases}
$$

$$
\begin{aligned}
: A &< \frac{T_{macro}+sh/\,ln\,(u/g)}{\eta C S_{micro}} \\
: \frac{T_{macro}+sh/ln\,(u/g)}{\eta C S_{micro}} < A &< \frac{T_{macro}+sh/\,ln\,(u/g)}{\eta C S_{micro}e^{-d\sqrt{S_{area}/\pi}}} \quad (27)\\
: A &> \frac{T_{macro}+sh/\,ln\,(u/g)}{\eta C S_{micro}e^{-d\sqrt{S_{area}/\pi}}}
\end{aligned}
$$

5 Results and Traffic Offload Guideline

5.1 Required Number of Base Stations

Numerical calculation is executed based on the model and process in chapter 4. Results are shown in Figure 6.

Figure 6(a) and (b) show number of required macrocell and microcell base stations n_{macro}, n_{micro} vs traffic per users B_u. In the figures, each base station cost ratio C defined by Equation (25) is changed from 5 to 100. As is seen in the figures, traffic offload is inevitable around user traffic of 1GB to 10GB/month.

For the general use of calculated results in any city, the traffic per user T_u [bps/user] is translated to the monthly data size per user B_u [byte/month/user], the traffic per unit area T_a [bps/m^2] and the monthly data size per unit area B_a [byte/month/m^2]. Following equations are translation equations.

$$
B_u = T_u \times 60sec \times 60min \times 24hour \times 30days/8bit = 3.24 \times 10^5 T_u \quad (28)
$$

$$
T_a = T_u N_d R_m \frac{R_d}{S_{area}} = 0.016T_u = 4.94 \times 10^{-8} B_u \quad (29)
$$

$$
B_a = B_u N_d R_m \frac{R_d}{S_{area}} = 0.016\,B_u = 0.016 \times 3.24 \times 10^5 T_u
$$

$$
= 5184 T_u = 3.24 T_a \quad (30)
$$

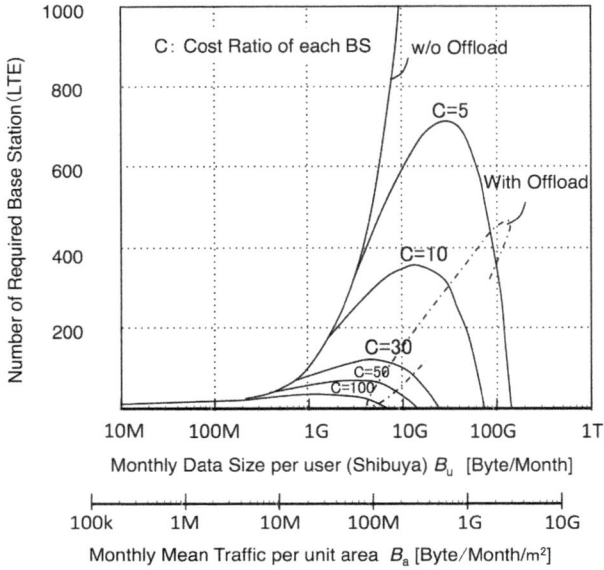

Figure 6   (a) Number of required mobile base stations for macrocell

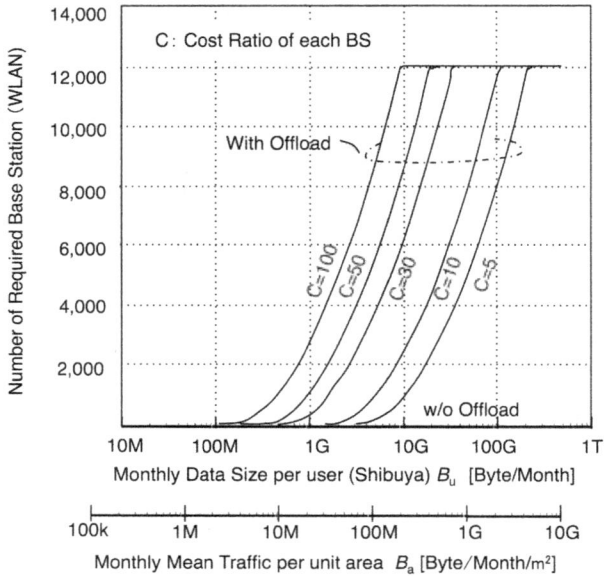

Figure 6   (b) Number of required mobile base stations for microcell

where N_d, R_m, R_d and S_{area} are daytime population, subscription ratio of mobile service, share of a specific mobile carrier and area of Shibuya ward, respectively, and N_d = 559,000, R_m =78%, R_d= 53.7%, S_{area} =15,110,000 m^2, as shown in Table 1.

5.2 Traffic Offload Ratio in Shibuya-Ward and Generalized Guideline for Various City

The required traffic offload ratio is calculated in Shibuya-ward, and the results is shown in Figure 7 according with the monthly data size per user B_u. As is seen in Figure 1, traffic will reach around 10GB/user/month in 5 years. In case of macrocell RBS cost is high, i.e. C=100, more than 80% of traffic should be offloaded to microcell base station, in the congested area. Even if the macrocell RBS cost becomes lower, i.e. C=5, several % of traffic should be offloaded in the high population area.

From this results, the required 50%-traffic offload year in Shibuya-ward will be calculated 2014 incase of C=100, and 2018 in case of C=5. Relation between the offload ratio δ [%] and the microcell installed area ratio α [%] is

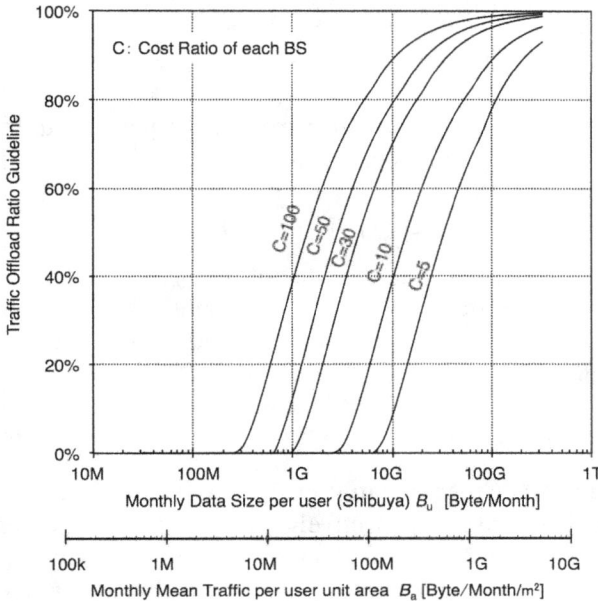

Figure 7   Traffic offload ratio guideline

shown in Figure 4, and from this figure, 50% offload means the microcell area ratio is 30% in case of the population concentration ratio $\beta = 3$.

If the traffic offload ratio, calculated in Shibuya-ward, is translated to the monthly data size per unit area B_a, the results obtained in Shibuya-ward is generally applicable to the other city in the world. In Figure 7, the monthly data size per unit area B_a is shown simultaneously. To follow the guideline, it is important that the parameter Monthly traffic per unit area B_a [byte/month/m^2] is easy to measure and transparent. The parameter B_a[byte/month/m^2] and traffic offload ratio may be measured by the mobile service operators already, however, it is not transparent to us, and all the congested cities in the world can not open such a data. Then, it is required to estimate the parameter by using the social and open bigdata. Next Section 7 shows the examples of the estimated results.

6 Required 50%-Traffic Offloading Year of Various Cities in the World

In this section, it is calculated that the several examples of the required 50%-Traffic Offloading year of various cities in the world. Required offloading year is briefly mentioned in the previous section. In this section, details of the calculation methods are defined. The calculation method is the following steps.

Monthly data traffic in n years after from 2012 per one user $B_{u,n}$ is calculated by the following equation.

$$B_{u,n} = B_{u,2012}(1 + AGR_t)^n \tag{31}$$

where $B_{u,2012}$, and AGR_t are monthly data traffic in year 2012 per capita and annual growth rate in user traffic, respectively. Monthly mean data traffic per unit area in n years after from 2012, that is denoted by $B_{a,n}$ is calculated by the following equation.

$$B_{a,n} = B_{u,2012}(1 + AGR_t)^n P_{ud,2012}(1 + AGR_u)^n \tag{32}$$

where $P_{ud,2012}$, and AGR_u are mean user density in year 2012 and annual growth rate in number of users, respectively. In this paper, mean user density $P_{ud,2012}$, is assumed and calculated from the population density P_d. For the easiness of estimation, it is estimated by $P_{ud,2012} = P_d$ in case of $AGR_u < 1.5$ %and $P_{ud,2012} = 0.8 \times P_d$ in case of $AGR_u > 1.5$ %.

Table 4 Guideline value of monthly traffic per unit area $B_a|_{x\%}$

Offload Ratio x [%]	Guideline of Monthly x%-Traffic Offload Per Unit Area, $B_a|x\%$ [MB/ Month/m^2]			
	C=100	C=50	C=10	C=5
10%	5	14	64	160
50%	24	45	208	480
90%	176	320	1,920	3,360

C : Cost ratio between Macrocell BS and Microcell BS

From Figure 7, we can obtain the guideline value of monthly traffic per unit area $B_a|_{x\%}$[MB/month/m^2]. Table 4 shows this value.

Using this guideline $B_a|_{x\%}$, required year n for offload is obtained from the following equation.

$$n = log((B_a|_{x\%} \times R_b)/(B_{u,2012} \times P_d))/log((1 + AGR_t)(1 + AGR_u)) \tag{33}$$

where, R_b denotes the bandwidth allocation ratio for various countries. There are wide variety of band assignments depend on the country, however 60MHz pair bands or 120MHz TDD bands are assumed in this analyses for the calculation easiness. Population density P_d is referred from Wikipedia [10]. The monthly data traffic in year 2012 per capita $B_{u,2012}$, the annual growth rate in user traffic AGR_t, and the annual growth rate in number of users AGR_u are referred from Cisco [6]. Data is summarized in Table 5.

Calculated results using Equation (33) are shown in Figure 8 (a) and (b). In the figures, base station cost ratio C is selected as variable parameter.

Figure 8 (a) shows the required 50%-traffic offloading year at the RBS cost ratio $C = 100$, and this shows that almost all cities of the high population density should be traffic offloaded by the end of 2014 or 2015. Figure 8(b) shows the required 50%-traffic offloading year at the RBS cost ratio $C = 5$, and this shows that almost all cities in high population density should be traffic offloaded by the end of 2016, even if the RBS cost of macrocell is reduced and increase the number of macrocell RBS.

7 Draft Proposal of Question

From the results of this research, Draft proposal of Question is proposed, as follow.

DRAFT NEW PROPOSAL OF QUESTION ITU-R XXX
Guideline of traffic offload ratio for mobile service
(2012-201x)
The ITU Radio Communication Assembly, considering

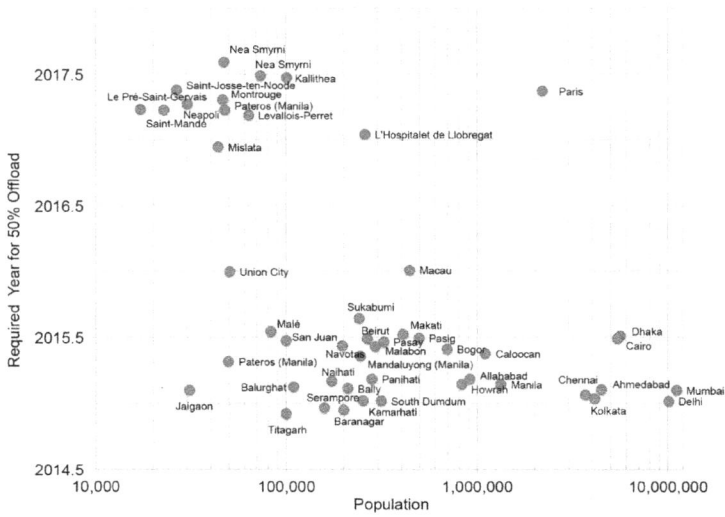

Figure 8   (a) Required year for 50% traffic offload (C=100)

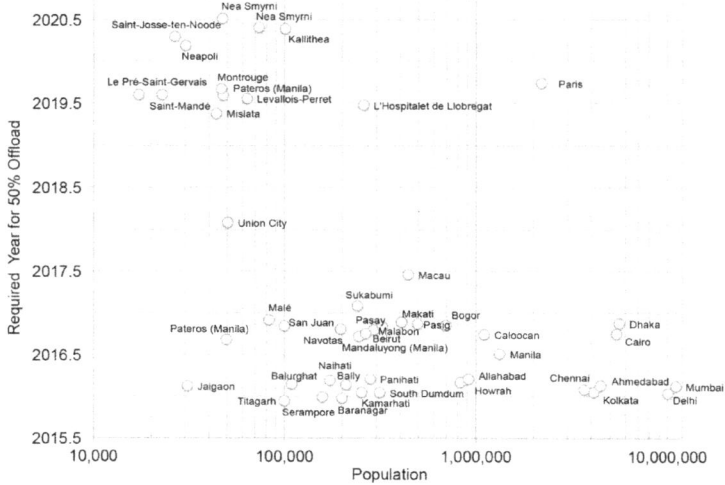

Figure 8   (b) Required year for 50% traffic offload (C=5)

Table 5 Population density and other parameters of various congested cities

City or Ward Name	Population	Estimated Subscribers	Area (km^2)	Population Density (/km^2)	Country	Traffic per User at 2012 Bu (MB)	AGR on Traffic AGRt (%)	AGR on Users AGRn (%)
Manila	1,660,714	1,328,571	38.55	43,079	Philippines	108	71	4.2
Bogor	866,034	692,827	21.56	40,169	Indonesia	93	63	3.9
Titagarh	124,213	99,370	3.24	38,337	India	23	108	7.9
Baranagar	248,466	198,773	7.12	35,220	India	23	108	7.9
Serarrpore	197,857	158,286	5.88	33,649	India	23	108	7.9
Pateros	61,940	49,552	2.1	29,495	Philippines	108	71	4.2
Delhi	12,565,901	10,052,721	431	29,155	India	23	108	7.9
South Dumdum	392,444	313,955	13.54	28,984	India	23	108	7.9
Kamarhati	314,507	251,606	10.96	28,696	India	23	108	7.9
Kolkata	5,138,208	4,110,566	185	27,774	India	23	108	7.9
Ahmedabad	5,570,585	4,456,468	464	22,473	India	23	108	7.9
Mandaluyong	305,576	244,461	11.26	27,138	Philippines	108	71	4.2
Levallois-Perret	63,225	63,225	2.42	26,126	France	244	47	1.4
Neapoli	30,279	30,279	1.17	25,879	Greece	771	39	1
Caloocan	1,378,856	1,103,085	53.34	25,850	Philippines	108	71	4.2
Chennai	4,616,639	3,693,311	181.04	25,501	India	23	108	7.9
Vincennes	47,372	47,372	1.9	24,802	France	244	47	1.4
Sukabumi	300,694	240,555	12.15	24,748	Indonesia	93	63	3.9
Saint-Mandé	22,737	22,737	0.9	24,714	France	244	47	1.4

(continued)

Table 5 Continued

City or Ward Name	Population	Estimated Subscribers	Area (km^2)	Population Density (/km^2)	Country	Traffic per User at 2012 Bu (MB)	AGR on Traffic AGRt (%)	AGR on Users AGRn (%)
Le Pré- Saint-Gervais	17,244	17,244	0.7	24,635	France	244	47	1.4
Saint- Josse-ten- Noode	26,488	26,488	1.14	23,235	Belgium	771	39	1
Malabon	363,681	290,945	15.76	23,076	Philippines	108	71	4.2
Mumbai	13,830,884	11,064,707	603	22,937	India	23	108	7.9
Jaigaon	38,689	30,951	1.69	22,893	India	23	108	7.9
Navotas	245,344	196,275	10.77	22,780	Philippines	108	71	4.2
Mont rouge	46,500	46,500	2.1	22,464	France	244	47	1.4
Banupur	11,647	9,318	0.52	22,398	India	23	108	7.9
Guttenburg	11,176	8,941	0.628	22,052	USA	763	55	1.7
Bally	260,906	208,725	11.81	22,092	India	23	108	7.9
Balurghat	135,737	108,590	6.37	21,309	India	23	108	7.9
Mislata	43,756	43,756	2.06	21,241	Spain	483	42	1.4
Pasay	403,064	322,451	19	21,214	Philippines	108	71	4.2
Kallithea	100,050	100,050	4.75	21,068	Greece	771	39	1
Paris	2,193,031	2,193,031	105.4	20,741	France	244	47	1.4
San Juan	124,187	99,350	5.94	20,907	Philippines	108	71	4.2

Nea Smyrni	73,090	73,090	3.52	20,740	Greece	771	39	1
Pasig	617,301	493,841	31	19,913	Philippines	108	71	4.2
Howrah	1,034,372	827,498	51.74	19,992	India	23	108	7.9
Dhaka	7,000,940	5,600,752	360	19,447	Bangladesh	108	71	4.2
Union City	62,715	50,172	3.29	19,066	United States	763	55	1.7
L'Hospitalet de Llobregat	257,038	257,038	13.62	18,872	Spain	483	42	1.4
Makati	510,383	408,306	27.36	18,654	Philippines	108	71	4.2
Naihati	215,303	172,242	11.55	18,641	India	23	108	7.9
Saint- Giles	46,931	46,931	2.52	18,623	Belgium	771	39	1
Macau	552,500	442,000	29.7	18,600	Macau	63	63	3.8
Cairo	6,758,581	5,406,865	374	18,071	Egypt	62	68	5.4
Beirut	331,366	265,093	20	18,063	Lebanon	62	68	5.4
Allahabad	1,142,722	914,178	63.38	18,030	India	23	108	7.9
Panihati	348,438	278,750	19.4	17,961	India	23	108	7.9
Malé	103,693	82,954	5.798	17,884	Maldives	108	71	4.2

Reference: http://en.wikiped.org/wiki/list_of_cities_Proper_by_Population_density.

http://www.cisco.com/web/solution/sp/vni_mobile_forecast_highlight/index.html, Accessed 20 june 2013.

a) that radio spectrum is the most important resource;
b) that smart phone is penetrating rapidly as social universal service;
c) that flat rate tariff is preferable to care of network neutrality and social utility;
d) that traffic offload to microcell can mitigate the requirements for new radio spectrum allocations;

decides that the following guideline be established

1. What types of traffic offloading technology is preferable?
2. What is the relevant evaluation parameter for offloading to minimize spectrum resource and investment cost of radio base stations?
3. What is the preferable measure of offload characteristics, that is transparent and observable including bigdata?

8 Conclusion

For the solution of heavy traffic concentration problem in wireless cloud era, the traffic offloading to microcell and the offload guideline is calculated at Shibuya-ward of Tokyo metropolitan, as a high traffic concentrating area in Japan. Guideline is based on the cost minimization strategy of the radio base stations and limitation of existing spectrum resource. Calculated results show that the traffic offloading to microcell is necessary in 5 years, even if the LTE is implemented at all radio base stations. Obtained results are translated to the universal parameters for the global use, and applied to various congested cities in the world. Required offload years are calculated for the high population cities and wards in the world, and results show that the traffic offload should be done very near future, even if the cost down of macrocell mobile radio base stations are done. From the proposed guideline, it is required to obtain the parameter of monthly traffic per unit area. The parameter can be directly obtained from the bigdata analyses from the operating mobile carriers, and it may be better to establish monitoring system and methodology. It is recommended that the parameters are opened to users and regulatory body of radio spectrum, for the enhancement of social utility. Another indirect method is the estimation from the social and open bigdata analyses, and the method is applicable for the region in that mobile service operation carrier has no data. At the same time, the urgent and real implementation of offloading technology including wireless agents, and the offloading promotion are required now.

Acknowledgment

I am indebted to Prof. Dr. Katsutoshi Tsukamoto, Osaka Institute of Technology and Associate Professor Dr. Takeshi Higashino, Nara Institute of Technology for their valuable discussions and comments, and also would like to thank Mr. Nobuyuki Shutto for their supports and research works on this issue in Osaka University.

References

[1] Sankaran, C. B., "Data offloading techniques in 3GPP Rel-10 networks: A tutorial," Communications Magazine, IEEE, Vol.50, No.6, pp.46–53, January 2012

[2] Singh Sarabjot, Dhillon S. Harpreet, Andrews G. Jeffrey, " Offloading in Heterogeneous Networks: Modeling, Analysis, and Design Insights," Trans. on Wireless Communications, IEEE, Vol. 12, No. 5, pp. 2484–2497, May 2013

[3] Aijaz A., Aghvami, H., Amani, M., "A survey on mobile data offloading: technical and business perspectives," Wireless Communications, IEEE, Vol. 20, No. 2, pp. 104–112, April 2013

[4] Ministry of Internal Affairs and Communications, "Report of present Communication Traffic in Japan; 2009," http://www.google.co.jp/url?sa =t&rct=j&q=&esrc=s&source=web&cd=1&ved=0CC4QFjAA&url= http%3A%2F%2Fwww.soumu.go.jp%2Fjohotsusintokei%2Fwhite paper%2Fja%2Fh21%2Fexcel%2Fl4303110.xls&ei=aozCUdC1NcPZr Qes4YDoAg&usg=AFQjCNE802U170XjPHCi88GzPIOMNS6HSw& bvm=bv.48175248,d.bmk, Accessed 20 June 2013

[5] Ministry of Internal Affairs and Communications, "Report of Communication Traffic in Japan: 2012," http://www.soumu.go.jp/johotsusintokei/ field/tsuushin06.html, accessed 20 June 2013.

[6] Cisco Corp., "VNI Mobile Forecast Highlights, 2012–2017", http:// www.cisco.com/web/solutions/sp/vni/vni_mobile_forecast_highlight/ index.html, Accessed 20 June 2013.

[7] Colin Clark, "Urban Population Density", Journal of Royal Statistical Society, Series A, Vol. 114, pp.490–496, 1951

[8] Yuji Handa Nobuyuki Shutto Takeshi Higashino Katsutoshi Tsukamoto and Shozo Komaki, "A Proposal of the SLA for Consumers and Measurements of Utility Functions to Specify the SLA," Annual report of JSICR, March 2007.

[9] Masahiko Nabe, Kenichi Baba, Masayuki Murata, et.al., "WWW Traffic Analysis and Modeling for Internet Access Network Design (in Japanese)," IEICE Trans. B-1, Vol 80, No.6, p.p. 428–437, June,1997.
[10] Wiki pedia, "List of cities proper by population density", http://en.wikipedia.org/wiki/List_of_cities_proper_by_population_density, Accessed 20 June 2013.

Biographies

Shozo Komaki, Professor, MJIIT, UTM. He was born in 1947. He received BS, MS and PhD degrees from Osaka University in 1970, 1972 and 1983, respectively. He joined to NTT Electrical Communication Labs. in 1972, where he was engaged in R&D on digital microwave radio systems. From 1990, he moved to Osaka University and engaging in the research on Radio on Fiber Networks, Wireless service over IP networks, Software Definable Radio Networks and Radio Agents. He is currently a Professor of Malaysia-Japan International Institute of Technology (MJIIT), Universiti Teknologi Malaysia (UTM).

Naoki Ohshima, Associate Professor, MJIIT, UTM. He was born in 1964. He graduated from Doctoral course of Department of Crystalline Materials

Science, Graduate School of Engineering, Nagoya University in 1993, and earned his Doctor degree of Engineering. His first faculty position was at Electrical and Electronic Information Engineering, Toyohashi University of Technology from 1993 to 1999. From 1999 to 2005, he belonged to Department of Advanced Materials Science and Engineering, Yamaguchi University. In 2005, he moved to Graduate School of Innovation & Technology Management, Yamaguchi University. He is currently an Associate Professor of Malaysia-Japan International Institute of Technology (MJIIT), Universiti Teknologi Malaysia (UTM).

Hassan Keshavarz Ph.D Candidate, MJIIT, UTM He was born in 1982. He received his M.Sc. at Computer Science in 2013 from University of Malaya (UM), and he got his BSc (Software Engineering) in 2007 in Iran. His main research interests include Big Data, Data Mining, Flow Analysis, and Internet of Things. He is a member of IEEE society and it can be communicated via hassan@mjiit.com. He is currently a Ph.D. candidate and research assistant in Malaysia-Japan International Institute of Technology (MJIIT) in the Management of Technology (MOT) Department at the Universiti Technologi Malaysia, Malaysia.

www.ingramcontent.com/pod-product-compliance
Lightning Source LLC
Chambersburg PA
CBHW061840220326
41599CB00027B/5350